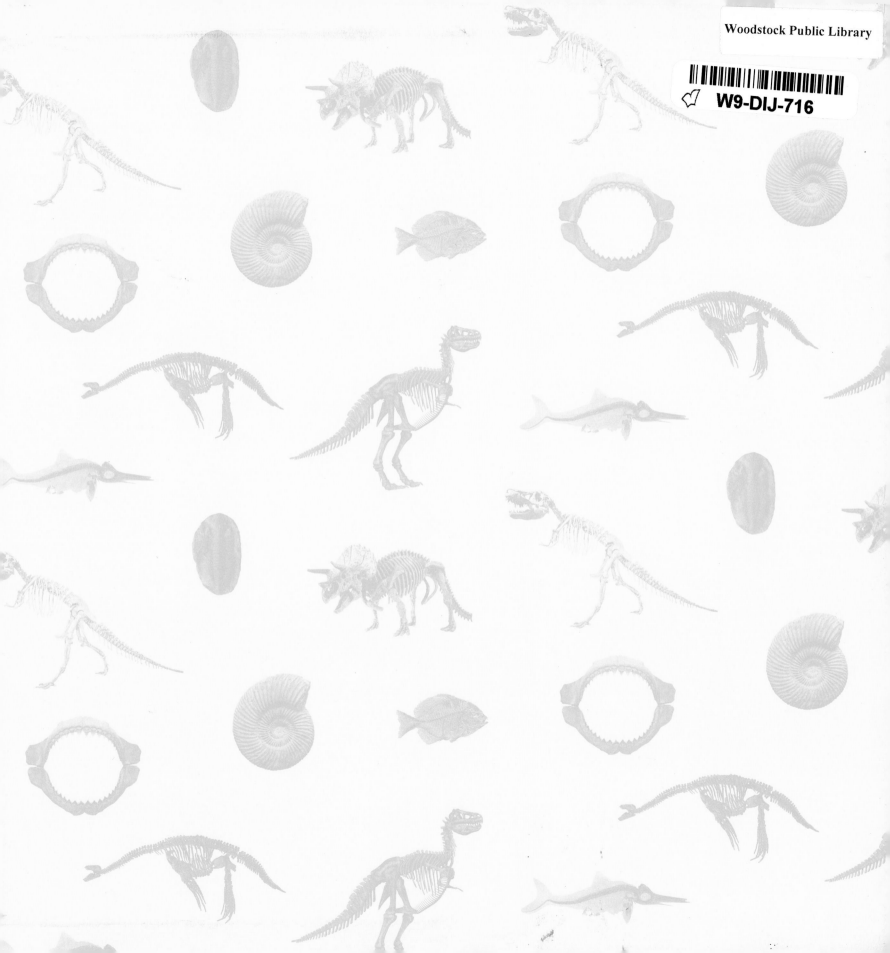

W9-DIJ-716

A CHILD'S INTRODUCTION TO

NATURAL HISTORY

A CHILD'S INTRODUCTION TO

NATURAL HISTORY

The Story of Our Living Earth—From Amazing Animals and Plants to Fascinating Fossils and Gems

By HEATHER ALEXANDER • Illustrated by MEREDITH HAMILTON

BLACK DOG
& LEVENTHAL
PUBLISHERS
NEW YORK

Black Dog & Leventhal Publishers
Hachette Book Group
1290 Avenue of the Americas
New York, NY 10104
www.hachettebookgroup.com
www.blackdogandleventhal.com

Printed in China

Cover design by Christopher Lin

Interior design by Sheila Hart Design, Inc.

APS

First Edition: June 2016

10 9 8 7 6 5 4 3 2 1

Black Dog & Leventhal Publishers is an imprint of Hachette Books, a division of Hachette Book Group. The Black Dog & Leventhal Publishers name and logo are trademarks of Hachette Book Group, Inc.

The Hachette Speakers Bureau provides a wide range of authors for speaking events. To find out more, go to www.HachetteSpeakersBureau.com or call (866) 376-6591.

The publisher is not responsible for websites (or their content) that are not owned by the publisher.

Library of Congress Cataloging-in-Publication Data is available upon request.

ISBN: 978-0-316-31136-6

DEDICATION

To my curious girl who loves sharks, coral reefs, cheetahs, seahorses, the Galapagos, and all things Pliny.
— *HA*

For Meg and Grace, who provide me with daily introductions to the natural world.
— *MH*

Contents

What Is Natural History?. 8

Let's Talk About Earth 10

Rocks, Minerals, and Gems 12

Try It: Hard Knock Life 14

When Did Life First Appear on Earth? 16

Once Upon a Time 19

Welcome to the Ice Age 22

A Lot of Time . . . a Lot of Change 24

Let's Talk About Plants 26

Microlife 30

Try It: Grow Mold 31

Let's Talk About Animals 32

Annelids. 34

Arthropods 35

Mollusks 42

Fish 44

Oceans and Coral Reefs 48

Amphibians 50

Freshwater 52

Reptiles 54

Deserts 58

Birds 60

Forests 64

Mammals 66

Grasslands 72

Tundra 74

Try It: The Blubber Glove 75

What's on the Menu? 76

Run and Hide: Camouflage and Other Defenses 78

I've Got Your Back: Partners and Parasites 80

Group Living 82

Stay, Go, or Sleep? 84

Animal Babies! 86

All Gone . . . and Going 88

All About Natural History Museums 92

Index 94

Size Comparisons 96

What Is Natural History?

Orangutans

You may have heard the term "natural history" in school or when you visited a museum. But what does it mean? What is this book about?

Let's break it down. **"Natural"** is something caused by nature. It's something that humans did not make. Do sour gummy candies come from nature? Is a cell phone part of nature? No. So what is?

Dolphin

Go outside and walk in the woods, kneel by a pond, swim in the ocean, dig in the snow or in the sand, or sit in your yard or in a park. What do you hear? What do you see? Nature is all around you—in the trees and grass, in birds in the air and the fish in the sea, in a squirrel on the ground, and in the tiny insect on your windowsill.

Ology Alert!

An "ology" is a subject that is studied. There are a lot of "ologies" in science. **Biology is the study of living things.**

Now, on to **"history."** History is defined as "knowledge gained by investigation." That means learning by looking closely at something. So in natural history, we are looking closely at nature. We are looking closely at everything that lives on the Earth—and understanding how they came to be, where they live, how they behave, and how they live together.

Sunflower

Dragonfly

Leopard cub

Guess what? You may already be a naturalist!

This book is an introduction to understanding all the mysteries and delights of nature. It's meant to inspire you to look more closely at the plants and animals around you and spark your interest in those that aren't nearby. Do you really like whales? Or Venus flytraps? Visit your library, bookstore, or nearby museum—there's so much more to learn and find out.

Another good reason to study natural history is to understand what the world has and what it has lost. You keep hearing about animals dying out and the planet changing. This book will introduce you to all that's important and unique in nature, and hopefully you'll agree that they're worth saving. You live on an amazing planet!

You don't need to be a scientist to study natural history. Anyone can be a **naturalist**. A naturalist is a person who studies nature. A naturalist is curious. She wants to know the names of the different kinds of sharks. He wants to see how a blue jay builds its nest. She wants to know why a snake sheds its skin. And why a spider doesn't get stuck in its own web.

What makes something "living"?

Living things on Earth have different sizes and shapes, but they all have some things in common. They all:

- breathe
- obtain and use energy
- grow
- react to light, heat, and water
- reproduce (have babies)

If something does only one of these things, it is not alive. It needs to do *all* of them.

Plants and animals are alive.

Poison dart frog

Get to Know: Pliny the Elder (23–79 CE)

Pliny the Elder was the first famous naturalist. He lived during the Roman Empire. He wrote an encyclopedia called *Natural History*, which contained 37 books. In them, he described the different kinds of honeybees, the different kinds of trees, how to get purple dye from a snail, and basically everything people knew back then about the natural world. Pliny was the first to organize all these facts and explain them in detail. For centuries, the books were used to teach science and medicine. Although many of his theories have been proven wrong today, Pliny's books got people excited about studying the world around them.

Let's Talk About Earth

Before we explore our amazing natural world, let's talk about Earth. We live on planet Earth. Earth is part of the **solar system**. The solar system includes the Sun, eight planets, all their moons, five dwarf planets (Pluto is one of them) and their moons, and billions of asteroids and comets. Some people call Earth the "third rock from the Sun." Do you know why?

Wow!
F A C T

Standing on Earth, the atmosphere (or sky) looks blue. If you travel into space, the sky would look black.

The Solar System

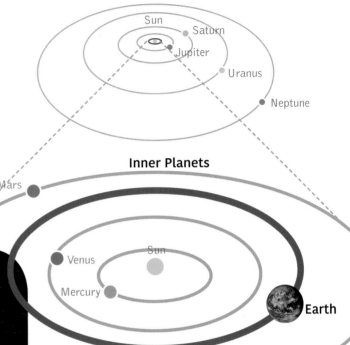

Outer Planets

Sun
Saturn
Jupiter
Uranus
Neptune

Inner Planets

Mars
Venus
Sun
Mercury
Earth

Earth is different from every other planet because:

Earth has a breathable **atmosphere**. The atmosphere is the layer of air around the Earth. The atmosphere acts like a blanket, protecting Earth from the intense heat of the Sun. Gravity keeps the atmosphere in place.

Air is made up mostly of the gas nitrogen, but it also contains some oxygen, argon, and carbon dioxide. When we breathe, we only use the oxygen. Without this combination of gases in the air, living things would die.

The Water Cycle

Earth does not have a never-ending supply of water. We use the same water over and over again.

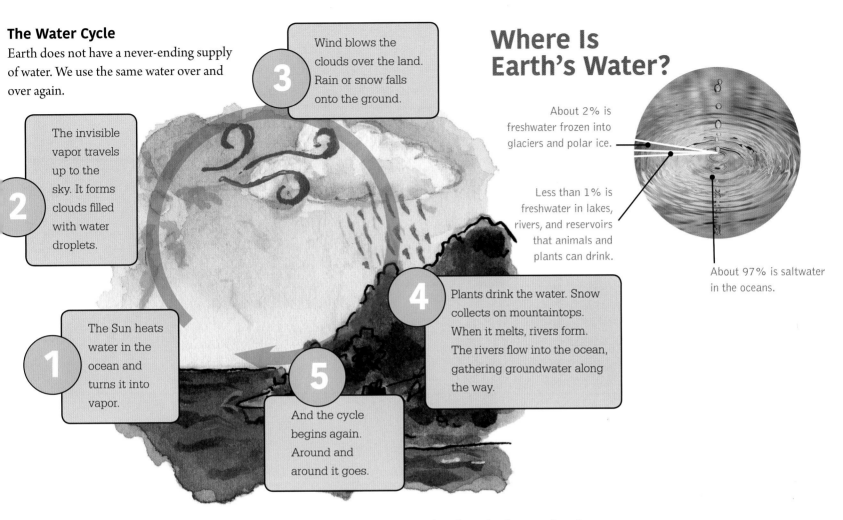

2 The invisible vapor travels up to the sky. It forms clouds filled with water droplets.

3 Wind blows the clouds over the land. Rain or snow falls onto the ground.

1 The Sun heats water in the ocean and turns it into vapor.

5 And the cycle begins again. Around and around it goes.

4 Plants drink the water. Snow collects on mountaintops. When it melts, rivers form. The rivers flow into the ocean, gathering groundwater along the way.

Where Is Earth's Water?

About 2% is freshwater frozen into glaciers and polar ice.

Less than 1% is freshwater in lakes, rivers, and reservoirs that animals and plants can drink.

About 97% is saltwater in the oceans.

Earth has **liquid water**. Earth is often called the "Ocean Planet." Water covers nearly 71% of the Earth's surface. Earth is the only planet in the solar system that has water as a solid (ice), a liquid (oceans, rivers, rain, etc.), and a gas (clouds and vapor). Liquid water keeps the natural world alive. It lets plants grow and animals drink. Oceans balance the temperature of the planet by absorbing the extra carbon dioxide in the air. Oceans also affect the weather by keeping cold and warm weather moving.

Healthy ocean = Healthy planet

Earth has **life**. Earth is the only planet in the solar system where life has been found, because there is air to breathe and water to drink. Earth's atmosphere keeps the planet at the perfect temperature to support life. Other planets are much hotter or much colder. A few other planets do have atmospheres and some water, but only Earth has the right combination.

Earth has **biodiversity**, meaning that *many* kinds of life are found here. There are over two million different living things, or **organisms**, on Earth—and those are only the ones scientists know about and have named. They believe there are millions more still to be found. Do you think you can discover a new plant or animal? If you do, you get to choose its name. One scientist named a slime mold beetle after Darth Vader, the *Star Wars* villain. Beyoncé has a horsefly named for her, and President Barack Obama has lichen named for him.

Rocks, Minerals, and Gems

Amber

Coal

Granite

Marble

Diorite

Lava Rock

Slate

Start by gathering some rocks and taking a look. At first, most appear gray or brown. Now look closer. Do you see some reddish and black rocks? Clear rocks? Rocks with chunks or stripes of color? A rock is made up of **minerals**, and minerals give rocks their color.

Some rocks are made from one mineral, such as silver or copper. But most are made of a mixture of minerals. Granite has many minerals. Do you have a granite countertop in your kitchen? If so, try to pick out all the different colors in the rock.

Coal and amber are rocks that are *not* made of minerals. Coal is plant material that has been squished together to form a solid. Amber is very old, hardened resin. Resin is a liquid from fir and pine tree bark, and sometimes bugs get trapped inside when it hardens.

Ology Alert!

Geology is the study of rocks, minerals, and the Earth.

Mineralogy is the study of minerals.

Gemology is the study of gems.

Salt crystals, called **halite**, are shaped like cubes.

Crystal Clear

Every mineral has a unique three-dimensional shape, called a **crystal**.

Gold is very soft and easy to hammer into different shapes.

Space Rocks

Minerals have been found in outer space, too. **Meteorites**, rocks that hit Earth from space, contain iron.

What a Gem!

Gems are pretty minerals. They look pretty, because their minerals bend the light. Once they are cut and polished, they are used for sparkly jewelry. Some large gems, such as diamonds and sapphires, cost a lot of money, because they are hard to find.

Heavy Metal

Gold, silver, and platinum are **precious metals**. They are found in thin layers in rocks.

Ores are rocks that have a lot of metals in them. The rock is crushed into a powder and the ore is taken out. Some ores are iron, copper, aluminum, and zinc.

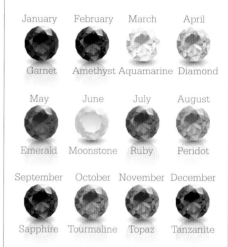
Surprise Inside

A **geode** looks like an ordinary rock on the outside, but inside it is filled with colorful crystals. A geode starts as a hollow rock. Mineral-rich water seeps inside. After thousands and thousands of years, crystals form inside.

Minerals All Around You

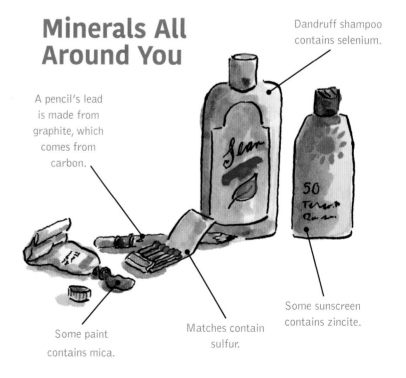

A pencil's lead is made from graphite, which comes from carbon.

Dandruff shampoo contains selenium.

Some paint contains mica.

Matches contain sulfur.

Some sunscreen contains zincite.

Mohs' Scale

In 1812, a German mineralogist named Friedrich Mohs created a scale to rank minerals based on their hardness. It is still used today. 1 is the softest, and 10 is the hardest. Each mineral can scratch only those below it on the scale (so a 2 can scratch a 1).

5.5
Steel nails are 5.5

10
DIAMOND
A diamond can cut steel. Only a diamond can cut another diamond.

9
CORUNDUM
Rubies and sapphires are types of colorful corundum.

8
TOPAZ
Used in jewelry

7
QUARTZ
Used to make glass, computers parts, and clocks

6
ORTHOCLASE FELDSPAR
Found in granite counter-tops and used to make glass and ceramics

5
APATITE
Found in bones and teeth

Wow! FACT

Most natural diamonds are between 1 to 3 billion years old!

Try It : Hard Knock Life

You need:

Three different rocks from outside
Paper and pencil
Your fingernail
A penny
A steel nail

To do:

1. Draw this chart on a piece of paper.

	Fingernail	Penny	Steel Nail	Hardness #
Rock #1				
Rock #2				
Rock #3				

2. Scratch Rock #1 with your fingernail. If you see the scratch, write YES in the Fingernail column. If you don't see a scratch, write NO.

3. Scratch Rock #1 with a penny. If you see the scratch, write YES in the Penny column. If you don't see it, write NO.

4. Scratch Rock #1 with a steel nail. If you see the scratch, write YES in the Steel Nail column. If you don't see it, write NO.

5. Now try to figure out the number of your rock's hardness using the Mohs' scale.

6. Repeat the above for Rock #2 and then Rock #3.

4	**3**	**2**	**1**
FLUORITE	CALCITE	GYPSUM	TALC
Used in toothpaste to prevent cavities	Used in cement	Used in crayons	Used in talcum powder

2.5
Fingernails are 2.5

3.5
Pennies are 3.5

Limestone can become marble. Marble was used to build India's Taj Mahal.

Metamorphic

Far below the Earth's surface, heat and pressure can cause a rock's minerals to break apart. Try this: Rub the palms of your hands together really fast. Do you feel heat? Now squeeze your hands together hard. Do you feel the pressure? Heat and pressure change, or morph, the minerals into new minerals. This new rock is called **metamorphic**.

A Rocky Start

Rocks are placed into three different groups by the way they are formed.

Igneous

The word "igneous" comes from the Latin word for "fire." These rocks started deep inside the Earth as super-hot liquid, called **magma**. Magma is lighter in weight than the rock surrounding it, so it pushes up to the surface. At the surface, the magma cools and hardens. It becomes igneous rock. The Earth's crust is made from igneous rock.

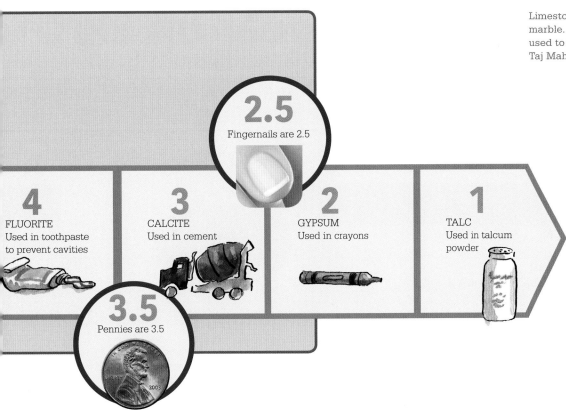

The faces on Mount Rushmore in South Dakota were carved from the igneous rock granite.

Sedimentary

Sediment is small pieces of rock that break off from larger rocks because of wind, rain, water, or ice. Some examples of sediment are sand, gravel, and clay. When sediment sticks together and hardens into thousands of layers, it forms sedimentary rock.

The Grand Canyon was formed after the Colorado River wore away layer after layer of sedimentary rock. This process is called **weathering** or **erosion**. Erosion is a slow process that changes the Earth.

When Did Life First Appear on Earth?

In the beginning . . . When exactly was the "beginning"? This isn't an easy question, but scientists believe the Earth formed about 4.5 billion years ago. How and when life began isn't known for sure, but scientists have come up with this timeline.

5 billion years ago

3 billion years ago
Bacteria began to use the energy of the Sun to make food. This produced oxygen in the air.

530 million years ago
Jawless fish appeared. They were the earliest animals to have backbones.

520 million years ago
Trilobites crawled all over the ocean floor. Trilobites are relatives of today's crabs, centipedes, and spiders. They had a hard outer covering and were one of the first creatures to have eyes. Some had eyes made up of thousands of lenses, each with six sides. Trilobites all died out, but many of their fossils remain.

490 million years ago
Life began to move from the oceans onto land. Algae and other plants began to grow. Small crustaceans crawled on the ground.

248 million years ago
The first true dinosaurs appeared. There were also flying reptiles, such as the **pterosaur**.

300 million years ago
Amphibians started to lay eggs with leathery shells. This allowed animals to be born on dry land for the first time. Some amphibians began to turn into reptiles.

200 million years ago
The Earth's climate grew very warm, and dinosaurs grew even larger. The long-necked *Diplodocus* was 90 feet (27 m) long! That's the same length as two yellow school buses. Early mammals lived in the same times but they were small and rodent-like. They ate insects and scurried up trees or burrowed underground to stay safe.

4.5 billion years ago
(4,500,000,000—
That's a lot of
zeros!) Way back
then, Earth was a
super-hot rock with
poisonous gases and
many chemicals.

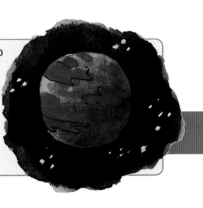

3.5 billion years ago
Earth cooled down, but the air didn't have oxygen, so organisms
couldn't survive. The oceans were filled with the minerals and
chemicals needed for life. This mixture
is called the **primordial soup**.
Life is believed to have started
in this "soup." Lightning bolts
may have caused a chemical
reaction that made the first
cell. A cell is the basic unit of
life, and all life on Earth came
from this first cell—you, your
dog, and the tree outside your
window. The first organisms were
single-celled bacteria.

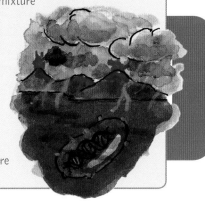

1 billion years ago
Because of the oxygen, multicellular
animals started to form. They stayed in
the ocean to avoid the strong radiation
from the Sun.

650 million years ago

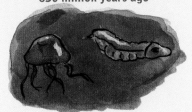

Soft worms and jellyfish-like animals
appeared in the ocean.

750 million years ago

**400 million
years ago**
Fish with jaws
appeared. The
enormous *Dunkleosteus* had razor-sharp
plates in its mouth and an
armor-like covering. Early
sharks swam in these seas.
They looked very much like
sharks look today.

380 million years ago
Very, very, very slowly,
fish began to develop
their fins into legs.
The four-legged fish
crawled onto land.
They became the first
amphibians and the
first vertebrates to live
on land.

360 million years ago
Trees and ferns grew large and tropical
forests spread. The amount of oxygen in
the air increased, and
animals began to grow
bigger, too. Some land
animals began to fly.
The biggest dragonfly
had a wingspan of 30
inches (75 cm)!

150 million years ago
The first bird evolved. The
Archaeopteryx looked like
a dinosaur, but it had
feathers, wings, and could fly.
It also had teeth.

130 million years ago
The first flowering
plants appeared.
Flowering plants
allowed many animals
to survive.

100 million years ago

75 million years ago

65 million years ago
The dinosaurs all went **extinct**. Scientists aren't totally sure why, but they think an enormous meteorite may have smashed into the Earth and created a great dust cloud. The cloud blocked the Sun. Without the Sun's rays, plants died and the climate turned too cold for the dinosaurs to survive. They also think that, around the same time, huge volcanoes may have let a lot of carbon dioxide and ash into the air. This also blocked the Sun.

60 million years ago
Mammals started to grow bigger. Why? The huge dinosaurs that had always hunted them were gone. Mammals no longer had to compete to eat the plants they needed to grow.

Extinct: When every single one of an animal or plant species has died out.

5 million years ago

1.8 million years ago
Ice Age mammals, such as the woolly mammoth, roamed Earth.

2.8 million years ago
Early humans, called **hominins**, took their first upright steps. Hominins looked a lot like apes, but hominins always walked on two legs. Hominins were shorter than modern humans, and they had longer arms.

195,000 years ago
Modern humans, called **Homo sapiens** (meaning you and me), first appeared in Africa. They had larger and more complex brains than other animals. **"Homo sapiens"** means "wise man" in Latin.

Present day

HOW DO WE KNOW ALL THIS? HUMANS WEREN'T ALIVE BACK THEN! The answer is: **Fossils**. When you hear "fossils," you probably think of dinosaur bones. But fossils aren't only about dinosaurs. Fossils form when minerals replace all or part of a dead animal or plant. Fossils are usually found in sedimentary rock. Most fossils were made when an animal died near water and was covered with mud. Some were made when animals froze to death or became trapped in tar pits. Fossils give scientists clues about life long ago.

Once Upon a Time...

What Is a Dinosaur?

Dinosaurs were animals that lived more than 230 million years ago and died out about 65 million years ago. That's a long, long time ago. Modern humans haven't even been alive for 200,000 years. Dinosaurs came in all shapes and sizes. Some were as tall as a six-story building. But some were only the size of a chicken. Some ate plants and some ate other animals. Some walked on two legs, some walked on four legs—and some even flew. Some had horns and scaly skin and others had feathers.

"Dinosaur" comes from a Greek word meaning "fearfully great lizard." Back then, "fearfully great" meant "awesome" or "amazing."

> DINOSAURS:
> • had long tails. The tails balanced the dinosaurs—the same way a seesaw with two equal-sized kids on either side stays balanced.
> • had scaly skin or feathers
> • had claws
> • had legs that stuck straight down from their bodies
> • lived on land

We now know that dinosaurs were not lizards. Why not? It's all in the legs. A dinosaur's legs reached straight down from its body—the way a horse's legs do. A lizard's legs stick out from the sides of its body.

Age of the Dinosaurs

Although it is often called the "Age of the Dinosaurs," not all dinosaurs lived at the same time. The *T. rex* and the velociraptor lived about 80 million years apart from each other.

The Plant Eaters

There were more plant-eating dinosaurs than meat-eaters. Plant-eating dinosaurs were usually larger in size than the meat-eaters. They often had long necks and very small heads. **Mamenchisaurus** had a neck that was more than half the size of its enormous body. Heavy, long-necked dinosaurs are called **sauropods**. Sauropods had to eat a lot of plants to fill up.

Plant-eating dinosaurs had to protect themselves from the ferocious meat-eaters.

The **Stegosaurus** had thick, bony plates on its back.

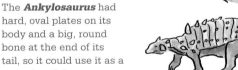

The **Triceratops** had three sharp horns on its head. Its name means "three-horned face."

The **Ankylosaurus** had hard, oval plates on its body and a big, round bone at the end of its tail, so it could use it as a swinging club.

Wow!
F A C T

Not all plant-eaters were big. The **Lesothosaurus** was the size of a chicken.

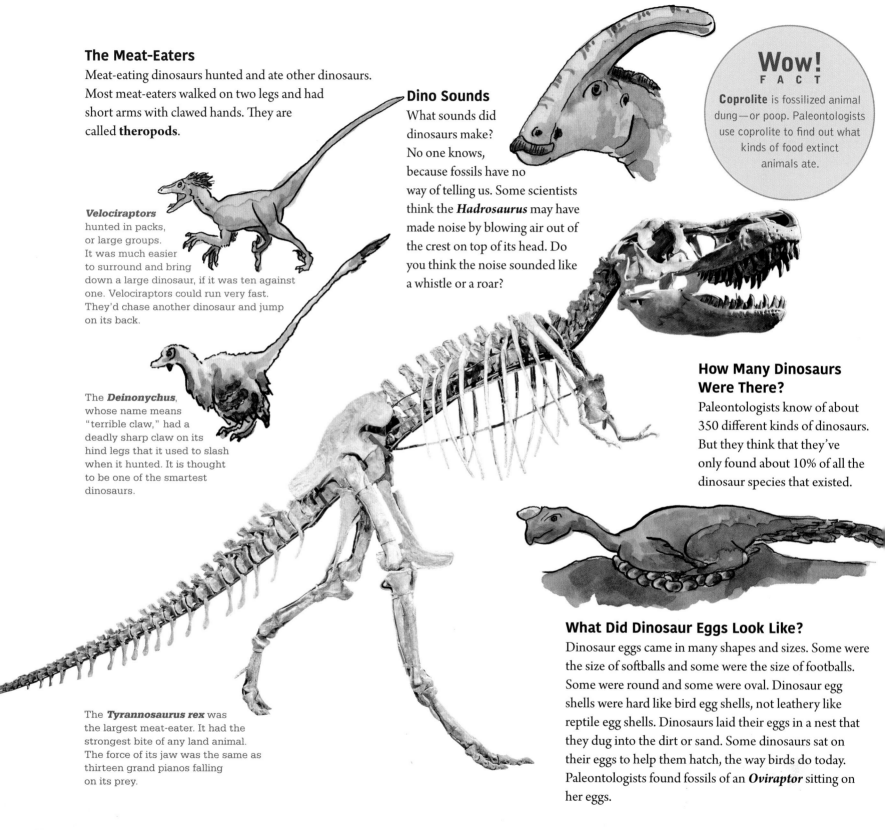

The Meat-Eaters

Meat-eating dinosaurs hunted and ate other dinosaurs. Most meat-eaters walked on two legs and had short arms with clawed hands. They are called **theropods**.

Velociraptors hunted in packs, or large groups. It was much easier to surround and bring down a large dinosaur, if it was ten against one. Velociraptors could run very fast. They'd chase another dinosaur and jump on its back.

The ***Deinonychus***, whose name means "terrible claw," had a deadly sharp claw on its hind legs that it used to slash when it hunted. It is thought to be one of the smartest dinosaurs.

The ***Tyrannosaurus rex*** was the largest meat-eater. It had the strongest bite of any land animal. The force of its jaw was the same as thirteen grand pianos falling on its prey.

Dino Sounds

What sounds did dinosaurs make? No one knows, because fossils have no way of telling us. Some scientists think the ***Hadrosaurus*** may have made noise by blowing air out of the crest on top of its head. Do you think the noise sounded like a whistle or a roar?

Wow! F A C T

Coprolite is fossilized animal dung — or poop. Paleontologists use coprolite to find out what kinds of food extinct animals ate.

How Many Dinosaurs Were There?

Paleontologists know of about 350 different kinds of dinosaurs. But they think that they've only found about 10% of all the dinosaur species that existed.

What Did Dinosaur Eggs Look Like?

Dinosaur eggs came in many shapes and sizes. Some were the size of softballs and some were the size of footballs. Some were round and some were oval. Dinosaur egg shells were hard like bird egg shells, not leathery like reptile egg shells. Dinosaurs laid their eggs in a nest that they dug into the dirt or sand. Some dinosaurs sat on their eggs to help them hatch, the way birds do today. Paleontologists found fossils of an ***Oviraptor*** sitting on her eggs.

Other Creatures During Dino Times

Sea reptiles and flying reptiles were not dinosaurs, but they were related to the dinosaurs—kind of like "dino-cousins."

The **Ichthyosaurus**, which means "fish lizard," had a fish-like body and huge eyes. It was a strong swimmer but had to come up to the surface to breathe.

Even though no dinosaurs swam in the ocean, that doesn't mean the waters were empty. Giant swimming reptiles with teeth filled the oceans, as well as jellyfish, squid, coral, and sharks.

No true dinosaur flew, but some prehistoric reptiles (not birds!), called **pterosaurs**, could fly. Pterosaurs had wings made from skin, not feathers. Some pterosaurs were the size of a blue jay and some were the size of an airplane!

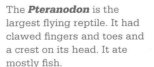

The **Pteranodon** is the largest flying reptile. It had clawed fingers and toes and a crest on its head. It ate mostly fish.

Get to Know: Mary Anning
(1799–1847)

Mary Anning has been called the world's greatest fossil hunter. She lived near the sea in Dorset, England. Her family was very poor, and she helped her father collect shells and fossils to sell in a stall on the beach. The famous tongue twister: "She sells seashells by the seashore" was written about Mary. As she wandered the rocky shoreline, Mary discovered small fossils of sea creatures that had lived a long, long time ago. Although Mary never went to school, she read a lot. She taught herself about rocks and animals. When she was 12 years old, Mary's older brother spotted what looked like a tooth in the rocks. It wasn't just a tooth. It was a skull. Mary went back with her hammer and carefully uncovered the full skeleton of a creature that looked like a crocodile! She'd found the first fossil of an *Ichthyosaurus*. Mary fossil-hunted for the rest of her life and discovered a *Plesiosaur*, a giant sea reptile, and many other prehistoric fish and reptiles. Famous scientists from all over the world visited her. Mary's fossils helped them understand how life had evolved.

Welcome to the Ice Age

The Ice Age lasted from about 1.8 million years ago until about 10,000 years ago.

Do you know why it was named the Ice Age? You guessed it—the Earth was covered with sheets of ice and snow. The animals that lived during the Ice Age were largely fur-covered, because they needed to keep warm. Many animals that we see today, such as cattle, deer, rabbits, and bears, roamed Earth back then. But there were also huge and amazing animals that that have since gone extinct.

Why Did Large Ice Age Animals Go Extinct?

Paleontologists believe one or both of these two things happened:

1. The climate became extremely cold. The frigid weather meant less food, and the large animals starved to death.

2. Humans appeared. When humans arrived in North America about 13,000 years ago, they hunted all the giant mammals for food. Most of these mega-mammals went extinct about 10,000 years ago.

Different Name, Same Beasts

Ice Age animals are sometimes also called prehistoric animals. "Prehistoric" means before history—or before *human* history started being written down.

Wow! FACT

In 2013, Yevgeny Salinder, an 11-year-old boy from Russia, was out walking his dogs when he stumbled upon a full woolly mammoth skeleton. Scientists informally named it "Zhenya," after Yevgeny's nickname. As worldwide temperatures rise and ice melts, more fossils and woolly mammoth tusks are being uncovered by regular people. Maybe you will find one too!

We're Going on a Mammoth Hunt!

When Thomas Jefferson was president, he thought woolly mammoths were still alive. In 1804, he sent explorers Lewis and Clark west of the Mississippi River, and one of their jobs was to search for woolly mammoths. They never found any. Why? Because they'd died out thousands of years earlier. This was the first time that many people realized that animals could go extinct.

Mammoth skeleton at the American Museum of Natural History in New York.

The **ground sloth** was an extra-large relative of a modern-day sloth. How large? Most were the size of an ox. Unlike today's sloths that spend their days up in the trees, some ground sloths stayed on the ground. They ate plants and stood on their hind legs to reach the tops of trees.

The **American mastodon** looked a lot like the woolly mammoth, except it was smaller (closer to the size of today's elephants). It also had cone-shaped teeth to eat leaves off the tops of spruce and pine trees and straight tusks.

Woolly mammoths were huge, hairy mammals that had two long, curved tusks, possibly used for fighting, a sloping back, and flat teeth for chewing. Their long trunks had two finger-like projections at the tip to hold objects. They ate grass and low shrubs. Woolly mammoths had a thick undercoat that kept them warm. Most woolly mammoths became extinct 10,000 years ago, but some survived on Wrangel Island in the Arctic Ocean until about 4,000 years ago.

Unlike many animals, **Homo sapiens** (otherwise known as humans) survived through the Ice Age. How did a tiny person kill a huge Ice Age mammal? People were smart. They knew they were no match in strength for the big, hairy animals, so they set traps. They also used spears with stone tips and darts.

The **dire wolf** was the largest canine that ever lived. Wolves and dogs are part of the canine family. The dire wolf was about 25% bigger than the biggest dog alive today. Its strong bite could crush its prey's bones. Dire wolf and saber-toothed cat fossils have been dug up in the same place, meaning they lived in the same area. Both were fearsome hunters. If a dire wolf fought a saber-toothed cat, which do you think would win?

The **saber-toothed cat** was about the size of a lion. It had two long, very sharp teeth to hunt and kill other animals. It could even kill a mastodon! Because it had short, powerful legs, it probably would sneak up and surprise-attack its prey instead of chasing it down.

The **giant short-faced bear** had a huge head and was bigger than any bear that ever lived. When standing, this enormous bear was 12 feet (4 m) tall! One swipe of its paw could kill another animal. And it could run 40 miles per hour (64 kph)—that's as fast as a coyote!

A Lot of Time
. . . a Lot of Change

Have you ever wondered why an animal or plant looks or acts the way it does? Why do giraffes have such long necks? Why are grasshoppers green?

The answers all have to do with **evolution.**

Get to Know: Charles Darwin (1809–1882)

Charles Darwin was a scientist who came up with the "theory of evolution." As a boy, Darwin loved collecting seashells, insects, coins—basically anything that could be collected. He studied at the University of Cambridge in England, and in 1831, he became a scientist and naturalist on a ship called the *Beagle*. The *Beagle* left from England and spent five years sailing around the world. When the *Beagle* anchored near land, Darwin collected different plants and animals. He drew pictures of them in his journals and wrote observations.

In 1835, the *Beagle* spent five weeks in the Galapagos, a group of islands off the coast of South America. The islands had been created by volcano lava. Some islands were very old and some were newer. Some had soil and trees and some did not. Darwin discovered a variety of animals that lived here but nowhere else on Earth. He also noticed that on each island the little birds, called finches, had different-shaped beaks. He counted 14 **species** of finches. Why were there so many different-looking finches in such a small group of islands?

Central America

Pacific Ocean

600 miles
(1,000 kilometers)

South America

The Galapagos Islands

Darwin returned to England and studied all he had found. He decided that all the finches had **descended**, or come from, two finches that had probably been blown over in a storm from nearby South America. These birds' children, grandchildren, and great-grandchildren settled on different islands. Each island grew different plants and had different animals. Darwin concluded that by living on different islands and eating different foods, the finches' beaks had slowly changed, or evolved. Some finches had small beaks to eat seeds. Some had sharp, pointy beaks to catch insects. Some had curved beaks to eat fruit.

Evolution is change in a species over time. Evolution is so incredibly slow that you can't see it happen. Let's say a skin color change begins today in a population of frogs that lives in a pond near your house. That change—maybe from light green skin to a darker green—would not be seen by you in your lifetime (or your kids or your grandkids, assuming your family stayed in the same house and the frogs stayed in the same pond). Evolution takes hundreds of thousands of years to be seen.

What Exactly Is Natural Selection?

Natural selection is the idea that the animals and plants best suited for their environment have an easier time surviving. They often have an **adaptation**, or a change in a body part or behavior. For example, the change in beak shape in Darwin's finches is an adaptation.

What's a Species?

A group of very similar animals is called a **species**. A **rooster** and a **duck** are both birds, but they are different species. Members of the same species usually look alike, act alike, and can have babies together.

So if the only food on one island were nuts with a tough shell to crack, the finches whose beaks were shaped in the right way to crack the shell got to eat. These finches grew strong, while the finches that couldn't crack the shell went hungry, grew weak, and died. The finches with the nut-cracking beaks then mated with other finches with nut-cracking beaks and had babies. Many of the babies also had nut-cracking beaks.

The population of nut-cracking-beaked finches then grew bigger and bigger. Darwin called this **natural selection**. In 1859, Darwin published a book called *On the Origin of Species*, in which he explained how natural selection was responsible for how every organism looks and acts.

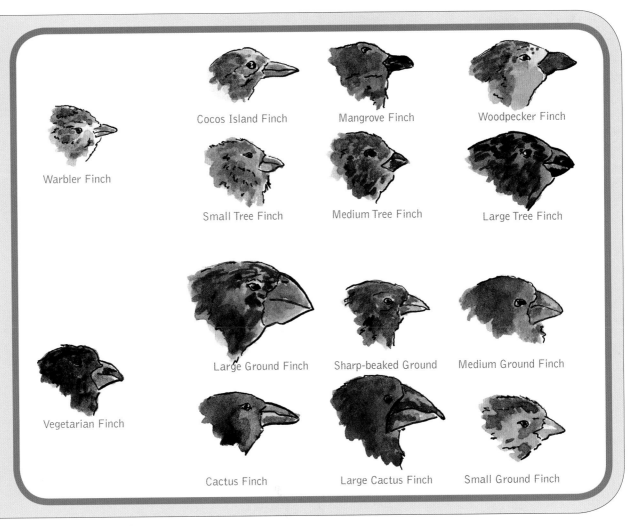

Warbler Finch

Cocos Island Finch

Mangrove Finch

Woodpecker Finch

Small Tree Finch

Medium Tree Finch

Large Tree Finch

Vegetarian Finch

Large Ground Finch

Sharp-beaked Ground

Medium Ground Finch

Cactus Finch

Large Cactus Finch

Small Ground Finch

Let's Talk About Plants

Plants are alive, the same way you are alive.
Plants eat and drink water. Plants breathe. They grow and react. They reproduce. So what makes a plant different than an animal?

What's the Difference?

PLANTS	ANIMALS
Can make their own food	Cannot make their own food
Cannot move around/have roots	Can move around freely
Breathe in carbon dioxide, breathe out oxygen	Breathe in oxygen, breathe out carbon dioxide

Ology Alert!

Botany (not exactly an ology, but close enough) is the study of plants.

What Do Plants Need to Grow?

Water, food, air, space, and sunlight

A Plant Starts in Many Different Ways

A daisy grows from a **seed**.

A daffodil grows from a **bulb**.

A potato grows from a **tuber**.

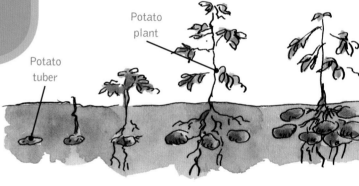

Potato plant

Potato tuber

Why Are Plants So Important?

- They make oxygen that animals, including humans, need to breathe. They help keep the air clean.

- They provide food that humans and animals eat. How many plants did you eat today?

- Their roots help hold the soil in place when it rains. Without plant roots, mudslides would wash away homes and hurt animals.

- They provide medicine. Aloe juice from the aloe vera plant helps take the sting out of sunburn. Digitalis, from the leaves of the foxglove, helps people with heart problems.

- They give us materials to make clothing, furniture, tires, and paper.

- They are homes for many animals.

The **burrowing owl** lives in the trunk of a tree.

Seed to Plant

Everything needed to grow a new plant is waiting inside a seed or a bulb. When a seed finds a good place (usually in soil), the right temperature (not too hot or too cold), and enough water, its hard outer covering, called a **seed coat**, breaks open. A shoot pushes out and begins to grow, or **germinate**. Its stem and leaves reach upward toward the sunlight.

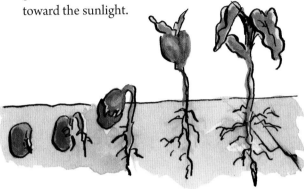

What Is Pollination?

Flowering plants contain a sticky yellow powder called **pollen**. In order to make new seeds, pollen from one plant needs to get to another plant.

Plants are divided into two groups: flowering and non-flowering.

Non-flowering— **Ferns** and **mosses** do not have flowers. They grow from **spores** instead of seeds.

Flowering— Each part of a flowering plant does a special job.

Flowers contain the male and female parts of the plant, and together they produce seeds.

Leaves help make food for the plant. Tiny holes in the leaves, called **stomata**, let out water vapor and let in carbon dioxide, a gas in the air. A green material in the leaves called **chlorophyll** uses sunlight to turn the carbon dioxide into a sugary food for the plant. This process is called **photosynthesis**. Photosynthesis is:

Carbon dioxide + water + sunlight = oxygen + glucose (sugar)

The **stem** supports the plant. Inside the stem, tubes called **xylem** carry water and minerals to other parts of the plant.

Roots anchor a plant to the ground, so it doesn't topple over. The roots also act like a straw and suck up water and minerals from the soil that the plant needs to grow.

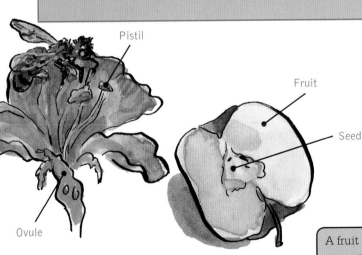

1 A flower's sweet scent and colorful petals attract a messenger, such as a bee. The bee drinks the sugary nectar inside a flower. While it is drinking, pollen rubs off on its legs.

2 The bee then carries the pollen with it when it flies to another flower to find more nectar. The pollen from the first flower falls onto the second flower.

3 It travels down the pistil to the egg, or ovule. Here a new seed begins to form.

4 A fruit forms around the seed. The fruit protects the seeds inside.

Pistil

Ovule

Fruit

Seed

A what? A **drupe** is a fruit that has a pulpy middle, surrounded by a shell or hard skin, and a seed or pit inside. They are also called **stone fruits**. Some drupes are: coconut, mango, olive, date, almond, apricot, cherry, nectarine, peach, and plum.

Apricot

Wow!
F A C T
A strawberry has about 200 seeds. It's the only fruit that has its seeds on the outside.

What Is Soil?
Soil is rock that has been crushed into many tiny pieces. Soil also has water, air, broken-up dead plants and animals (called **organic matter**), and lots of living creatures, too.

What Is a Weed?
A weed is a plant that is not wanted. Weeds often grow very fast, crowding out other plants, drinking up the water in the soil, and blocking the sunlight. Their seeds spread quickly.

A **dandelion** is a weed.

On the Move
A plant's fruit not only protects its seeds, but also helps to scatter them. Think about this: If a plant dropped its seeds directly into the soil underneath it, new plants would sprout all around it. The original plant would have to share the soil's water and minerals with all its "babies." There wouldn't be enough to go around, so some plants would die. Not a good plan. Enter the hungry animal!

The animal is attracted to sweet, juicy fruit. It gobbles up the fruit and the seeds, and then takes a walk. Later, it poops out the seeds. Or a bird carries a lot of berries as it flies, or a squirrel holds many nuts as it climbs through the trees, and some drop onto the ground. Whatever the way, the seeds now take root in new soil.

Spice It Up
Many of the foods you eat are flavored with herbs and spices. An **herb** comes from the leaf of a plant. Basil, oregano, mint, and parsley are all herbs. A **spice** comes from the seed, stem, bark, root, or berry. Cinnamon, nutmeg, and ginger are all spices.

Cinnamon

Leaves Come in Many Different Shapes

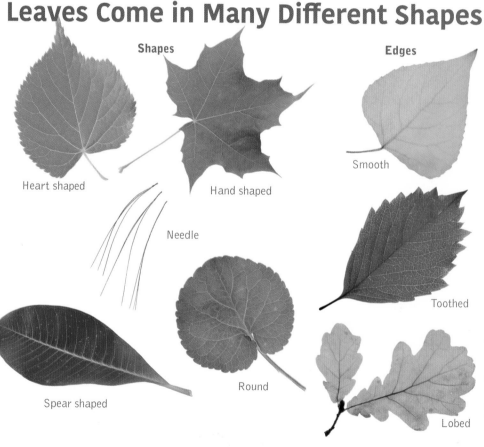

Shapes

Edges

Heart shaped

Hand shaped

Needle

Smooth

Toothed

Spear shaped

Round

Lobed

Don't Touch Me

Plants need to protect themselves from hungry and thirsty animals. But how do they do that if they can't run away?

Ouch! Some plants grow their own sharp weapons. The leaves on a **holly** bush have sharp points to stop animals from taking a nibble. **Rose** bushes and **lemon** trees have thorns. **Cactuses** and **pine** trees have spines and needles.

Disguise. Some plants pretend to look like something else. The **lithops** of Africa has gray-brown leaves that look like stones. Thirsty animals don't try to drink water from a stone!

Danger! Some plants have toxins or poisons that can make animals sick—or even kill them. **Poison ivy** has a substance called urushiol on its surface that causes a painful, itchy rash for humans. One way to spot poison ivy is to remember the rhyme:

Leaves of three, let it be;
berries white, poisonous sight.

Name Game

Many plants were named after what they look like.

Snapdragon flowers look like a dragon. If you squeeze the sides, the dragon's mouth opens and closes.

The **bird of paradise** looks like a colorful tropical bird.

Meat-eating Plants

Plants that eat insects and even small frogs are called **carnivorous plants**. These plants usually live in places where the soil doesn't have enough nutrients, so they need to get tricky and trap their food.

The **Venus flytrap** creates its own prison. An insect touches the tiny hairs on the edges of its leaves, and the leaves snap close, trapping the insect inside. The plant then digests the insect.

Trees

Trees are the biggest plants. The **trunk** is a tree's thick stem, and it is covered by **bark**. Bark protects the insides of the tree, the same way your fingernails protect the tips of your fingers. Each year a different layer of growth makes a trunk fatter and creates a ring. If a tree is cut down, you can see rings in the wood. Count the rings and you will know how old the tree is.

HALL OF FAME

Largest flowering plant: The ***Rafflesia arnoldii*** can grow 3 feet (1 m) across and weigh up to 24 pounds (11 kg). It is found in Indonesia's rainforest, and its flowers smell like rotting meat! Flies are attracted to the bad smell and pollinate the flowers.

Smallest flowering plant: The **watermeal** is smaller than one sprinkle on a cupcake!

Fastest growing plant: **Bamboo** can grow 35 inches (89 cm) in one day.

Heaviest seed: The **coco de mer palm** grows a seed that can weigh up to 50 pounds (22.7 kg) —that's the same as five medium-sized bowling balls!

Rafflesia arnoldii

Microlife

Right now, right where you are sitting, you are surrounded by a trillion living things. What's wrong? You don't see them? Just because you don't see them, doesn't mean they aren't right next to you—or right on you.

Microorganisms or **microbes** are simple, teeny-tiny creatures that are so small that you can't see them without looking through a microscope. Microbes are found everywhere. They make up of most of the life on Earth. There are trillions and trillions of microbes just in the area around you!

Plant or Animal?

Every living organism is made up of supersmall cells. Humans are made up of hundreds of trillions of cells. Many microbes are made up of only one cell.

Microbes are alive, but they aren't plants and they aren't animals. Why? They have different cells than plants and animals do.

Wow! FACT

There are more microbes on your hand than there are people on the planet.

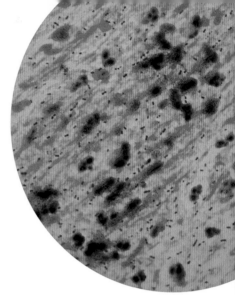
Dipplocci bacteria

Bacteria Everywhere

Bacteria are the simplest microbe. Thousands of bacteria could fit into the period at the end of this sentence. The air is filled with bacteria—and so is the soil in your yard, the food you eat, the toys you play with, and the chair you sit on. There are even bacteria inside you!

Tiny bacteria come in three different shapes—spiral, ball, and rod.

Some bacteria are helpful and some are harmful. The helpful

Spiral   Ball   Rod

bacteria help animals digest food, control the nitrogen in the air and create oxygen, and are used in medications. Harmful bacteria that cause diseases are often called **germs**. Less than 5% of all microbes are germs.

Salmonella is a kind of bacteria found on turtles, tortoises, and other animals. These bacteria don't hurt the turtle, but they can make people very sick. Salmonella is also found in some raw eggs. Always wash your hands with soap and warm water after you touch an animal or raw egg.

Ology Alert!

Microbiology is the study of microbes.
Mycology is the study of fungi.

Virus cell

Hermann's tortoise

Fungus Amongus

For many years, scientists thought **fungi** were plants. But fungi are more like animals. Fungi cannot make their own food. They get food from the soil or from breaking down dead material. Fungi can be found on old food in the refrigerator, on people's feet, and on the ground in a forest.

Mushrooms are fungi. Pigs and foxes eat mushrooms. So do people—but be careful. Some wild mushrooms are poisonous.

Wow!
F A C T

The honey mushroom in Oregon's Malheur National Forest is called "the humongous fungus." It covers an area of more than 3.4 square miles (9 km^2)! That's as big as 1,645 football fields. This monster mushroom grows underground and is thought to be thousands of years old.

Lichen looks like a plant, but it's not. It is a mixture of fungi and algae. Lichen does not have roots, so it can grow on rocks.

Try It : Grow Mold

Have you ever reached for a piece of bread and spotted a blue-green blob on it? That blob is **mold**. Mold is fungi. Mold grows where there is moisture. Mold can grow on rocks and walls, and it can also grow on food.

You need:
- 4 pieces of bread
- 4 sealable plastic bags
- Labeling marker
- Water

To do:
1. Place one piece of bread inside each plastic bag.
2. Sprinkle some water on the bread inside only two plastic bags. Label those bags to show which bags got water.
3. Leave the bags wide open for an hour to get air on the bread. Then seal all four bags tightly closed.
4. Place two bags (one with wet bread and one with dry bread) in a sunny place, such as on a windowsill.
5. Place the other two bags (one with wet bread and one with dry bread) inside the refrigerator.
6. Check the bags every day. Take notes or pictures to record what's happening. Which bags begin to grow mold first? Why?

Results (read only after finishing your experiment)
Mold needs moisture to grow, so your wet bread probably grew mold faster than your dry bread. Mold also grows faster in warm environments, so the bread left in the sun probably grew mold first. Cold stops fungi and bacteria from growing, which is why foods last longer in the refrigerator.

Get to Know: Antonie van Leeuwenhoek
(1632–1723)

Antonie van Leeuwenhoek was the first person to ever see a microbe. He used his own microscope. The lens of the microscope magnified what he was viewing, or made it look larger. His microscope was better than all others at that time, because it made something look 200 times larger. It has been said that one day, he scraped his teeth and put the gunk under his microscope. He saw tiny living things moving around. Then he scraped the gunk off the teeth of an old man who never brushed his teeth. There were tons of squirming creatures! He didn't know it then, but those creatures were bacteria. Bacteria in our mouths cause plaque, which can cause cavities. With the invention of the microscope, scientists have been able to study microbes and find cures for diseases.

Let's Talk About Animals

There are so many different kinds of animals—animals that fly, swim, lay eggs, have hair, or have feathers. How does a scientist make sense of them all when he or she sets out to study the natural world?

Scientists organized all animals into different groups called **classes**. The animals in a class are alike in either how they look or how they behave. Grouping them makes it easier to study them. There are many, many animal classes. This book focuses on eight main classes: **Annelids, Arthropods, Mollusks, Fish, Amphibians, Reptiles, Birds,** and **Mammals.**

Backbones

Animals are divided into **vertebrates** and **invertebrates.** "Vertebrate" means an animal has a skeleton with a backbone, or spine. You are a vertebrate. Mammals, birds, fish, reptiles, and amphibians are all vertebrates. Invertebrates do not have a backbone. Worms, jellyfish, insects, spiders, and crabs are invertebrates. Earth is home to many more invertebrates than vertebrates. Over 96% of all the animals on Earth are invertebrates.

Great white shark

Body Temperature

Every animal is either warm-blooded or cold-blooded.

Warm-blooded animals can make their own body heat. Whether it is very hot or very cold outside, warm-blooded animals have body temperatures that stay the same. Warm-blooded animals sweat and pant when they're hot and shiver when they're cold. You are a warm-blooded animal.

Cold-blooded animals have body temperatures that get hotter and colder depending on the temperature outside. Their bodies grow cold when the Sun sets at night, because the air around them loses its warmth. Their bodies become warmer when the Sun is out and their bodies can soak up the heat. A cold-blooded animal is active when it is warm. When it is cold, it doesn't move or eat much.

Pembroke Welsh Corgi

Sungazer Lizard

Predators and Prey

Animals can't make food the way plants can. Out in nature, animals must find their own food. Sometimes that food comes from plants. And sometimes that food is another animal. A big part of most animals' days—and nights—is keeping themselves from becoming someone else's dinner.

A **predator** is an animal that hunts other animals for food. The animal it hunts is called the **prey**. Not all predators are fierce, such as a shark hunting a seal. A ladybug hunting an aphid is also a predator. Predators aren't "bad guys"—they're just hunting so they can eat to stay alive.

The three main predator weapons are sharp **teeth, claws,** and **jaws.**

Home, Sweet Home (from Small to Big)

Every animal and plant has a home, and no animal or plant lives completely alone.

A **habitat** is an animal or plant's natural home. A habitat has everything that organism needs to survive. A habitat contains one or a few species.

An **ecosystem** is a community of all living things that live, feed, reproduce, and interact in the same area. An ecosystem is made of many smaller habitats. An ecosystem can be as small as a tree or as big as an enormous desert. An ecosystem contains a large number of species. In a healthy ecosystem, there are more animals of prey than predators. If it were the other way around, there would be nothing for the predators to eat.

A **biome** is a large region that contains many smaller ecosystems with similar climates. "Climate" and "weather" are not the same thing. **Weather** is what's happening outside your window right now. **Climate** is the pattern of weather over a long stretch of time. The amount of sunshine and rain tells you what kind of climate you are in.

There are land biomes (**grassland, tundra, desert, tropical rainforest, deciduous forest,** and **coniferous forest**) and aquatic biomes (**freshwater** and **marine** or **ocean**). We'll talk more about each biome throughout the book.

The lily pad is the dragonfly's habitat.

The pond is the ecosystem where the lily pad lives.

The pond is the part of the freshwater biome.

Annelids

Earthworm

ANNELIDS:
- are invertebrates
- are segmented
- have no legs or hard skeleton

"Slimy" and "squirmy" is how annelids are often described. Annelids are segmented worms. "Segmented" means that their tube-like bodies are divided into small sections. Annelids have no bones, but a tube filled with liquid helps their body keep its shape. Earthworms, tube worms, and leeches are annelids.

Ring Around the Worm

In Latin, "annelid" means "little ring." There is a ring around each segment of an annelid. Each segment is like a compartment with blood vessels, nerves, and muscles inside. Because of this, an annelid can go on living even if one or more segments are cut off or lost (except if it loses its head). Some earthworms can have hundreds of segments. Annelids move by contracting their segments.

HALL OF FAME

Longest earthworm: The **giant Gippsland earthworm** from Australia can grow to be up to 9.8 feet (3 m) long! That's about twice your height!

Largest leech: The **giant Amazon leech** is up to 18 inches (46 cm) long.

A **leech** has a leaf-shaped body with segments, but no rings. Its stretchy body expands when it is filled with liquid food. Some leeches suck blood.

Heads or Tails?

The easiest way to tell which end is which is to watch an earthworm burrow. It will usually move head-first.

Follow the Light

Most earthworms are reddish-brown in color, although some are pink, purple, or gray. Earthworms do not have eyes or ears, but they have light receptors on one end of their body that lets them know where they are.

Why Are Worms Slimy?

Worms let off a fluid that helps them breathe through their skin and lets them slide through the soil.

Nature's Plow

Earthworms are extremely important to the environment. They burrow through the soil. As they wriggle, they allow air to flow through the soil. When they dig, they eat the soil and dead plants and insects. They digest it and pass it out their back end, making the soil richer with nutrients.

Wow!
F A C T

If all the soil that has passed through earthworms' bodies was piled up, it would make a mountain five times higher than Mt. Everest.

Arthropods

Arthropods are a *huge* class of animals. How huge? About 85% of all known animals in the world are arthropods. **Insects, myriapods, crustaceans,** and **arachnids** are all arthropods. Arthropods are found in the air, on land, and in the water.

Dungeness crab

Ology Alert!

Arthropology is the study of arthropods.
Arachnology is the study of arachnids.
Carcinology is the study of crustaceans.

What Makes It an Arthropod?

1. An **exoskeleton**: Arthropods do not have backbones. They have a hard shell that covers and protects their entire body. Because the shell is stiff and doesn't stretch, an arthropod has difficulty growing. An arthropod must **molt**, or get rid of its old shell, so it can grow a new, bigger shell.

2. A **segmented** body: Arthropods have a body made up of more than one part, or segments. The segments link together like a chain. The segments allow arthropods to move their bodies, even though they have a shell.

3. **Many jointed legs**: Arthropod means "jointed leg." Joints allow a leg to bend. You have joints in your elbows and knees.

Arthropod	Number of Segments	Number of Legs
Spider	2	8
Ant	3	6
Lobster	21	10

Ants

A **green cicada** molts, getting rid of its old shell.

Crustaceans

"Crustacean" means "hard-shelled."
Lobsters, crabs, shrimp, and barnacles are all
crustaceans. They are often called "the insects of
the sea" because, unlike other arthropods, crustaceans have
gills and most live in the sea. However, some can walk on land.
Lobsters, crabs, and shrimp have pincers on their front two legs
to catch food and fight off predators. Crustaceans are **scavengers**.
Scavengers feed on dead plants or animals.

Wow!
F A C T

A **lobster** can be right-handed
or left-handed. How can you tell?
If it is right-handed, its right
claw will be larger. If it is
left-handed, its left claw will
be larger.

Some shrimp live in saltwater
and some in freshwater. The
freshwater **fairy shrimp**
always swims on its back.

A **hermit crab** does not
have its own hard shell, so it
squishes its body into a shell
that a snail has thrown away.
When it outgrows one shell,
it leaves it behind and looks
for a new one.

Myriapods

Myriapod means "many feet." Myriapods have
two body parts (a head and a long body), two antennae,
and eighteen or more legs. Myriapods live in damp soil
and under logs and rocks. Centipedes and millipedes are
the two main kinds of myriapods.

What's the Difference?

CENTIPEDE	MILLIPEDE
Long, flat bodies and long legs	Bodies are rounded on top and flat underneath with short legs.
Up to 354 legs	Up to 750 legs
One pair of legs on each body segment	Two pairs of legs on each body segment
Carnivores: They eat insects and worms. They have poisonous front claws to help them kill their prey. Their claws are too small to pierce human skin.	Herbivores: They eat plants and rotting material.
Fast	Slow

Australian Redback Spider

Arachnids

Many people think that spiders are insects. Wrong! Spiders are arachnids. Arachnids have eight legs, and insects have six legs. Other arachnids are scorpions, ticks, mites, and daddy longlegs. The word "arachnid" comes from an ancient Greek myth about a girl named Arachne. Arachne challenged the goddess Athena to a weaving contest. When Arachne lost, Athena turned her into a spider.

ARACHNIDS:
- have eight legs
- have two body parts
- have no antennae or wings

How Does a Spider Spin a Web?

All spiders can make silk. Not only is their silk waterproof, it's also stronger than a steel wire of the same thickness. A spider has a special tube called a **spinneret** at the end of its abdomen. To weave a web, a spider releases silk threads from its spinneret. Some threads are sticky, and some are non-sticky. Since the spider knows where it placed the non-sticky threads, it makes sure to walk along those, so it doesn't get trapped in its own web. The spider spins the threads in different patterns. Spiders are born knowing how to spin webs.

Orb Web  Sheet Web 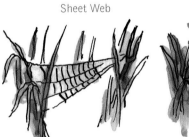 Funnel Web

Different spiders weave different-shaped webs.

How Does a Web Work?

BAM! A fly flies into the web. The threads vibrate. This alerts the spider—which has bad eyesight—that dinner has arrived. The spider wraps up the fly with sticky thread. Then it bites the fly with its poisonous fangs. The poison paralyzes the fly, making it unable to move. Spiders cannot chew or swallow, but the poison turns the fly's insides into a liquid. The spider then sucks out the liquid fly guts.

The hairy **tarantula** looks fierce, but most tarantula bites are harmless to humans. Tarantulas have retractable claws on their legs, similar to those on cats, so they can keep a strong grip while climbing.

Mites are the smallest arachnids. They are smaller than the period at the end of this sentence. Most mites are **parasites**, which means they feed on other animals. Some drink blood. Some burrow under skin. And some like to live in dust—especially under a bed or in a carpet.

HALL OF FAME

Largest crustacean: The **giant Japanese spider crab** is 12 feet (4 m) from claw to claw.

Smallest crustacean: The **water flea** is smaller than 1/100 inch (0.2 mm) long.

Largest arachnid: The **Goliath birdeater** is a **tarantula** that measures 11 inches (28 cm) across. It eats birds!

Largest myriapod: The **giant African millipede** can grow to 15 inches (38 cm) long.

Insects

Can you believe there are more than one million known species of insects? For every one human, there are 100 million insects! Insects are the only invertebrates that can fly.

Dragonfly

Ology Alert!

Entomology is the study of insects.

INSECTS:
• are arthropods
• are invertebrates with a hard exoskeleton
• are cold-blooded
• hatch from eggs

Is There a Difference Between an Insect and a Bug?

Yes! A bug is a kind of insect. A bug has a beak-like mouth with a hard straw that it uses to suck up juice from plants and animals. Not all insects have this. "True bugs," as they are called by scientists, include **aphids, cicadas, stink bugs, water bugs,** and **bed bugs**. So . . . *all bugs are insects, but not all insects are bugs.*

Aphid

Flies

The **housefly** is the most common fly. A housefly can eat only liquids. It spits its saliva onto food to make it liquid.

"Mosquito" means "little fly" in Spanish. Have you ever been bitten by a **mosquito**? Only the female mosquito bites mammals, which is what you are. The male mosquito eats plants. Many insects have a **proboscis**, which is a long, thin mouth part. A female mosquito has long, thin needles inside her proboscis. When she bites a mammal, she sticks the needles into the skin. She uses one needle to push her saliva into the skin to thin out the blood. Then she sucks it up through another needle. Mosquito saliva often makes the mammal's skin itchy and red. Some mosquitoes spread disease when they bite. Bats, birds, lizards, spiders, and frogs all eat mosquitoes.

Proboscis

Mosquito

All insects have:

Three main body parts: the head, the thorax (the middle part), and the abdomen (the end part).

Most insects have 1 or 2 sets of wings attached to the thorax.

Fly

Two antennae. These feelers on the head help an insect smell, touch, taste, and find prey.

Compound eyes. An insect's eyes are made up of thousands of very small eyes. Each small eye faces in a different direction. Together, all these tiny eyes give a great view to the front, back, and side. Compound eyes are good for seeing close-up movement. They are not good for seeing far away.

Tiny air holes called **spiracles** help insects breathe.

Six legs.

Beetles

There is only one species of human. There are 350,000 species of beetle! You can tell a beetle by its shell-like cover.

Ladybugs are beetles. Most ladybugs are round and either red or yellow with black spots. But some ladybugs are orange, gray, black, brown, or even pink. How many spots does a ladybug have? Some have 7, 12, or 22. The most spots ever counted were 24. The spots on a ladybug's back have nothing to do with its age. Despite the name, not all ladybugs are girls. Boy ladybugs are called ladybugs, too. In the United Kingdom and Australia, ladybugs are called ladybirds.

Fireflies are also beetles. They have a special organ in their stomachs where chemicals are mixed together to make a flashing light. This is called **bioluminescence**. A firefly uses its light to warn away predators and to talk with other fireflies. Each species of firefly has its own pattern of flashes, like a code, that lets them find one another.

Firefly Flash Patterns

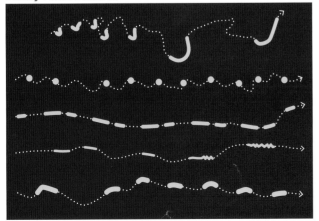

Ants

In an anthill, every ant has a job to do. The **queen** is the largest ant. She is the head of the family and the only one that lays eggs. The **worker ant** gathers food, cleans, and protects the anthill. The **drone** mates with the queen.

Get to Know: Charles Henry Turner
(1867–1923)

Charles Henry Turner was an entomologist, who was a pioneer in insect behavior studies. Born in Cincinnati, Ohio, to newly-freed slaves after the Civil War, he was the first African-American to earn a masters degree in biology from University of Cincinnati and the first to earn a Ph.D. in zoology from University of Chicago. But because of racism, Charles had a hard time getting a job as a college professor. Instead, he taught high school science in Missouri, while doing his own research and writing over 70 scientific papers. He was the first scientist to prove that insects can hear. He proved that honeybees can see color and recognize patterns. He also proved that cockroaches learn by trial and error. He built a maze for the cockroaches. He found that the cockroaches remembered when a path led to a dead end and did not take that path again. His important discoveries helped scientists better understand the insect world.

Bees

Honeybees and **bumblebees** live in hives that can have up to 60,000 bees.

Honeybee

How Do Honeybees Make Honey?

A bee sticks its long tongue inside the middle of a flower and sucks up the nectar. The nectar flows into a bag in the bee's stomach, called a honey bag, where it is turned into honey. Later, the bee spits the honey back out its mouth onto the honeycombs inside the hive. This honey is very watery, so the bees inside the hive fan it with their wings to evaporate the water and thicken it. Then the bees seal the honeycomb with wax to protect the sweet honey. They will eat the honey in the winter when there are no flowers or nectar. One bee only makes 1.5 teaspoons (8 ml) of honey its entire life. But since bees live in huge groups, together they produce a lot of honey.

Bee Bread

Bees use their jaws and tongue to scrape **pollen** from flowers. The pollen sticks to the tiny hairs on their body. The bees mix the pollen with the nectar or honey to make bee bread, a sticky substance that they feed to the younger bees.

Bee Sting

Many bees have a sharp stinger with pointed barbs on the end of their abdomen. When an animal gets too close, the bee pushes its stinger into it. However, when the bee moves away, the stinger stays in the skin, so the end of its abdomen pulls off and the bee dies.

Let's Dance

When bees find a really good spot to gather nectar from flowers, they fly back to the hive and do a "waggle dance." The angle of the dance based on the location of the Sun and the number of waggles they do tell the other bees what direction to go and how far to fly to find the nectar.

Ology Alert!

Apiology is the study of bees.

Butterflies

A butterfly is born as a caterpillar. It crawls and eats leaves. The caterpillar becomes a butterfly when it is an adult. Then it flies and drinks nectar.

From Caterpillar to Butterfly

There are four stages to a butterfly's life.

1 A female butterfly lays her eggs on the leaf of a plant.

2 A caterpillar hatches out of the egg. This is called the **larval stage**. The caterpillar spends most of its time eating leaves. The caterpillar grows fast. It gets too big for its exoskeleton and has to molt several times.

3 In a few weeks, it is ready to change into a butterfly. This change is called **metamorphosis**. The caterpillar attaches itself to a branch or leaf. It sheds its skin. A **chrysalis,** or hard shell, forms around the caterpillar. Inside, it is changing into a butterfly. This is the **pupal stage**.

4 Then the chrysalis cracks open. The butterfly comes out. This is the **adult stage**.

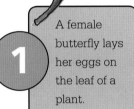
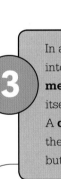

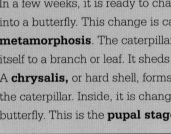

Green tiger beetle

Many Insects Are Powerful Jumpers

Stand in one place. Now jump forward. How far did you jump compared to your height? A **grasshopper** can jump 20 times the length of its body. That's like you jumping from home plate to first base in one bounce!

Mollusks

Mussels

MOLLUSKS:
- are invertebrates
- are cold-blooded
- have soft bodies
- must keep their bodies moist
- have a mantle

The word "mollusk" comes from a Latin word that means "soft." Mollusks are soft-bodied creatures that don't have skeletons. Mollusks are found everywhere—in oceans, lakes, rivers, and on land.

Many mollusks have an outer shell that grows along with it. A mollusk's thick shell protects it from enemies and from drying out. A mollusk must stay moist, or it will die. Have you ever collected shells on a beach? These shells were once the homes of mollusks. The shell is empty, because the dead mollusk's soft body was eaten by another animal.

Slowly They Go

Gastropods are the largest group of mollusks. The word "gastropod" means "stomach foot." Gastropods have one foot attached to their bellies, making them extremely slow. Snails and slugs are gastropods.

Gastropods that live in the ocean (**whelks** and **cone shells**) breathe through gills. Gastropods that live on land (**snails** and **slugs**) breathe through lungs.

Slug

Sea snails

Wow!
F A C T

An **ocean quahog clam** named Ming was pulled up from 262-foot (80-m)-deep waters off the coast of Iceland in 2006. Scientists counted the rings on its shell and calculated that Ming was born in 1499, making it 507 years old! When scientists pried open Ming's shell to study its insides, Ming died.

A mollusk's soft body is divided into three parts:

Mantle
A thin, skin-like covering that sometimes helps to form a shell.

Foot
Most mollusks have one muscular foot. The foot is used to move from place to place. It is also used to dig and anchor themselves in the sand or mud.

Head/Body
The soft, fleshy part above the foot contains the heart and other internal organs.

Close the Door!

Clams, oysters, mussels, and scallops are **bivalves**. Bivalves are mollusks that have two parts to their shell. The two halves look almost alike and are connected by a hinge. Bivalves clamp both halves closed when a predator draws near. Predators can't pry open the shell. Only the bivalve can open and close its shell from the inside.

Clam

Slugs and snails secrete slime when they crawl. Slime makes it easier for them to move and protects them from sharp rocks and twigs. A slug can crawl over an upright razor blade, pointed needles, or a sharp knife and not be cut!

I Want to Hold Your Hand, Hand, Hand . . . .

The name "**cephalopod**" means "head-foot." A cephalopod's limbs are attached to its head. Instead of a foot, cephalopods have at least eight **tentacles**, or arms, that move them through the water. Their arms are covered with powerful suckers to grab prey. Squid, octopuses, and cuttlefish are cephalopods.

Most cephalopods do not have a protective shell. Instead, they are fast and smart.

Giant Clam

Colossal Squid

Wow! FACT
A starfish can **regenerate**, which means it will grow a new leg if one is pulled off by a predator. If a starfish is cut into many pieces, it can live if it has kept part of its center.

Octopuses squirt a dark ink at whales, fish, and seals that try to eat them. They also can bite with sharp, hook-like beaks at the ends of their tentacles.

Star Power

Echinoderms are spiny sea creatures. Starfish, sea urchins, and sand dollars are echinoderms. Many echinoderms are shaped like a star or a wheel.

Echinoderms have no brain or eyes. They move along the sea floor using hundreds of tiny tube feet on the underside of their bodies.

Sea urchin

Big Blobs

Cnidarians are sea creatures with jelly-like bodies and stingers. Jellyfish, sea anemones, and corals are all cnidarians. They are the simplest of all animals. They have only one opening for eating food and passing out waste. They don't have eyes. They use their sense of smell and touch to find prey and stay away from predators. Cnidarians all sting to capture prey.

Jellyfish swim by opening and closing their bodies like umbrellas. This motion allows them to zoom up through the water.

Fish

FISH:
- are vertebrates
- lay eggs or give birth to live young
- live in either saltwater or freshwater
- have gills and fins

Fish are found in all types of water, both fresh and salty. Having gills and fins and living underwater makes fish different from other vertebrates. All fish are cold-blooded, except for a few such as the **opah**.

Opah

Why Don't Fish Sink?

Fish have a balloon-like organ called a **swim bladder**. The swim bladder gives the fish buoyancy, or the ability to float.

Swim bladder

How Do Fish Breathe?

All animals need oxygen to breathe. Most animals get their oxygen from the air—but not fish. Since fish live under the water, they need to get their oxygen from the water. As a fish swims, water passes through its gills. The gills take the oxygen out of the water and pump it into the fish's blood and organs. Fish need a steady supply of water over their gills to breathe.

The **mudskipper,** found in the swamps of Southeast Asia, can climb out of the water and survive on land for several hours. It has a pouch filled with water around its gills that it uses to stay moist.

Built to Swim

Have you ever walked through waist-deep water? Water is about 1,300 times heavier than air, and it pushes against you. Most fish are long and thin and pointed at both ends. This **streamlined shape** helps fish speed through the water. Airplanes also have a streamlined shape, but for moving through air.

Fish have FINS to help them swim.

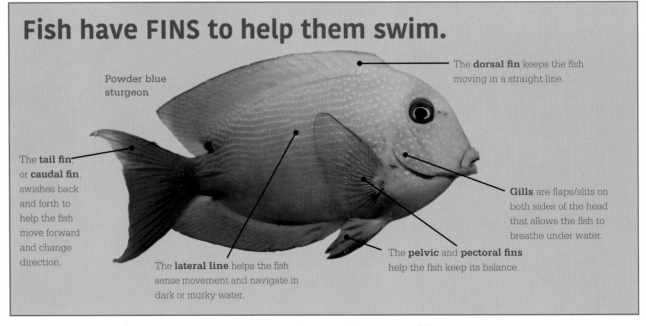

Powder blue sturgeon

The **dorsal fin** keeps the fish moving in a straight line.

The **tail fin,** or **caudal fin,** swishes back and forth to help the fish move forward and change direction.

The **lateral line** helps the fish sense movement and navigate in dark or murky water.

Gills are flaps/slits on both sides of the head that allows the fish to breathe under water.

The **pelvic** and **pectoral fins** help the fish keep its balance.

Scales

Tiny overlapping plates, called scales, cover many fish. Scales are waterproof and protect fish. Just like the rings on a tree trunk, the rings on a fish's scales can help you tell how old the fish is.

Scale Patterns

Bass

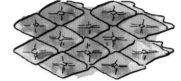

Salmon

Gar fish

Shark

There are three main groups of fish:

1 **Jawless** fish have no biting jaws. Instead, they have big holes for mouths. They have no scales and are long, slimy, and eel-like. **Lampreys** attach their round mouth to aquatic animals and suck their blood.

Lamprey

2 **Bony** fish have skeletons made from bones. Most fish, such as tuna, cod, and trout, are bony fish. They have smooth scales, and a swim bladder so they can float.

Blue fin tuna

3 **Cartilaginous** fish, such as sharks, skates, and rays have skeletons made from flexible cartilage, the same bendable material as in the tip of your nose or your ears. They have jaws with sharp teeth and rough skin. Cartilaginous fish don't have a swim bladder, so they must swim all the time.

Stingray

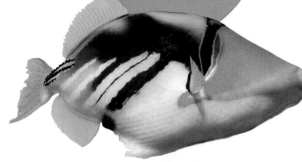

Wow!
FACT

Humuhumunukunuku apua'a is the longest fish name. It is the Hawaiian name for the reef triggerfish. The name is almost longer than the fish itself!

How Do Fish Sleep?

Fish do sleep, but not in the way mammals do. Fish sleep is more like when you daydream. They are not fully awake or fully asleep. They rest and slow down their heart rate. Some fish lie on the sea floor. Some tuck themselves behind rocks or coral. Some fish take a break from swimming and hover in place. But fish don't ever stop moving completely. They need to keep water flowing past their gills so they can get oxygen. And fish don't ever close their eyes. Why? Except for some sharks, fish don't have eyelids.

Down Deep

Some fish live in the deepest parts of the ocean where there is no sunlight. Many of these fish make their own light.

The **parrotfish** makes itself a sleeping bag when it wants to sleep. It secretes a jelly-like substance made of mucus and covers itself in it. This mucus sleeping bag stops predators, so the parrotfish can get some zzzzzs.

The **anglerfish** has a long flesh "rod" that dangles in front of its face. The end of the rod glows. Tiny fish are attracted to the light and then the anglerfish eats them!

HALL OF FAME

Biggest fish: The **whale shark** is 41 feet (12 m) long.

Heaviest fish: The **ocean sunfish** can weigh up to 2,200 pounds (1,000 kg).

Smallest fish: The **stout infantfish** is only 1/4 inch (8 mm) long.

Fastest fish: The **sailfish** swims 68 miles per hour (110 kph). This is faster than the speed most cars drive on a highway!

Longest living fish: The **rougheye rockfish** can live over 200 years!

Most poisonous fish: The **stonefish**, which lives at the bottom of the ocean, has spines that release deadly venom.

Stonefish

Whale shark

Hammerhead Shark

Shark Attack!

All sharks are predators, and they have several rows of sharp, triangular teeth ready to take a bite. As soon as a shark's tooth falls out or is broken, a new one from the row behind moves forward to replace it. A shark has an unlimited supply of teeth, and it will lose thousands in its lifetime.

The **great white shark**, the **hammerhead shark**, the **tiger shark**, the **mako shark**, and the **bull shark** are the most dangerous sharks. Their top-notch senses make them fierce predators. Sharks can smell one drop of blood in an area of water the size of an Olympic swimming pool! They can hear the sounds of wounded prey from miles away. They also have a strong sense of taste. Sharks will often take a "test bite" of their prey to see if they like the taste.

Great white shark

Get to Know: Dr. Eugenie Clark
(1922–2015)

Dr. Eugenie Clark was a famous ichthyologist who was often called the "Shark Lady." She founded Mote Marine Laboratory in Florida, a top center for shark research. As a child, she spent hours at the aquarium in New York City, wishing she could swim with the sharks. After studying zoology, Eugenie's wish came true. She became one of the first diving biologists at a time when few women had careers in marine biology. She was the first to prove that sharks have a memory and can be trained. She taught lemon sharks to hit a target with their snouts for food. She also discovered a fish called the Moses sole that produces a natural shark repellent. She dove into caves off the coast of Mexico to examine sharks that were suspended under water (local fishermen called them "sleeping sharks") and found that not all sharks had to move to breathe. And she once rode on the back of a 50-foot (15-m) whale shark! She was never bitten by a shark.

Oceans and Coral Reefs

If you could look down at Earth from outer space, what would you see? Blue. Oceans cover almost three-quarters of the Earth. An ocean is a huge body of saltwater. There are five different oceans: Pacific Ocean, Atlantic Ocean, Indian Ocean, Arctic Ocean, and Southern Ocean.

Did you know that you can sail around the world from sea to ocean to sea without ever stopping? That's because all the oceans of the world are connected.

Why Is the Ocean Salty?

Because it has salt in it. The same salt that you put on food is in the ocean. Most animals can't drink seawater, because the salt gets into their blood and their kidneys can't get rid of it without a lot of fresh water. But some ocean animals can drink it. Whales and seals have kidneys that can process the seawater. Penguins and gulls have a special gland in their bodies that removes the salt from their blood.

What's the Difference Between Oceans and Seas?

Seas are smaller areas of salty water that are partially or completely surrounded by land. There are 54 seas recognized by **oceanographers**, scientists who study the ocean.

Way Down Deep

The deepest parts of the ocean are called **trenches**. The Mariana Trench in the Pacific Ocean is the deepest place on Earth. It's about 36,000 feet (11,000 m) deep. How deep is that? The tallest mountain in the world, Mount Everest, could fit inside and there would still be 7,000 feet (2,133 m) of water on top!

Coral Reef

A **coral reef** is a huge, bustling, underwater community. It is home to about 30% of all marine life—that's 4,000 species of fish, 700 species of coral, and thousands of types of plant and animal life. There is more variety of species on a coral reef than almost anywhere in the world. Coral reefs are found in warm, shallow saltwater.

Plant or Animal?

Coral is actually a tiny animal, called a coral **polyp.** It has a soft body, stomach, and tentacles—kind of like a jellyfish. Polyps grow a hard outer skeleton. When the polyp dies, its skeleton is left behind. Another coral animal grows on top of the skeleton, and another, and another—until a coral reef is formed.

Coral eat **plankton**, which are floating microscopic plants and animals. They grab the plankton with their tentacles and put them into their mouth. Coral eat at night.

Coral get their color from tiny, colorful algae called **zooxanthellae** that live inside them. If they didn't have the algae, coral would all be white.

Three Kinds of Coral Reefs

FRINGE REEF—The reef is attached to the shore.

ATOLL—The reef is in a ring around an extinct volcano.

BARRIER REEF—The reef is separated from the shore by a channel of water.

= Coral

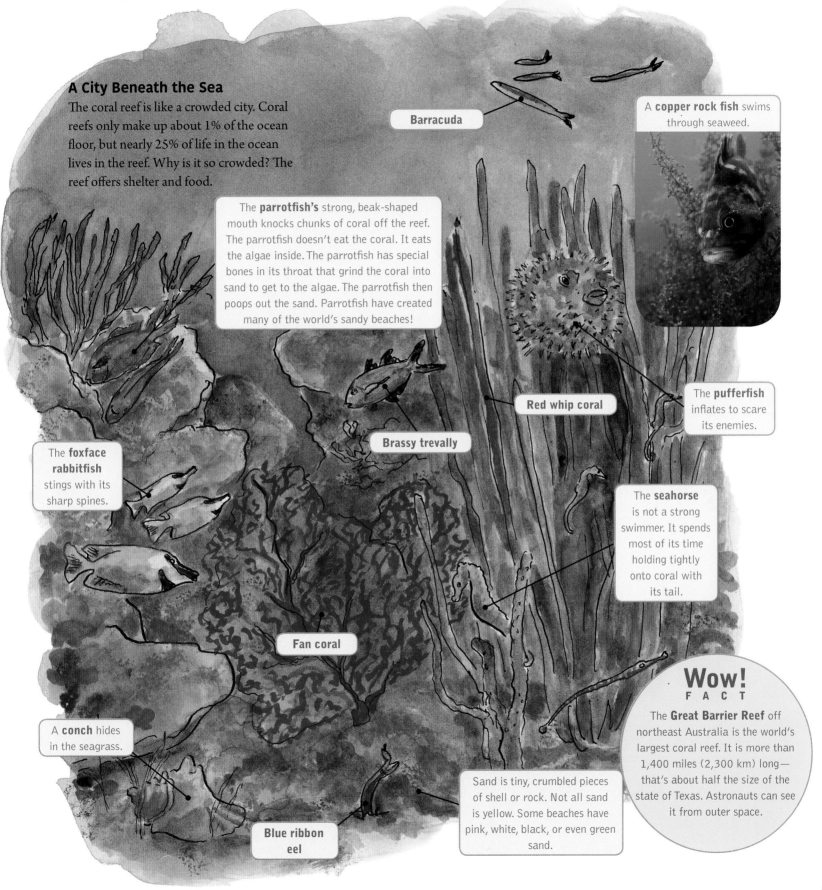

A City Beneath the Sea

The coral reef is like a crowded city. Coral reefs only make up about 1% of the ocean floor, but nearly 25% of life in the ocean lives in the reef. Why is it so crowded? The reef offers shelter and food.

Barracuda

A **copper rock fish** swims through seaweed.

The **parrotfish's** strong, beak-shaped mouth knocks chunks of coral off the reef. The parrotfish doesn't eat the coral. It eats the algae inside. The parrotfish has special bones in its throat that grind the coral into sand to get to the algae. The parrotfish then poops out the sand. Parrotfish have created many of the world's sandy beaches!

Red whip coral

The **pufferfish** inflates to scare its enemies.

Brassy trevally

The **foxface rabbitfish** stings with its sharp spines.

The **seahorse** is not a strong swimmer. It spends most of its time holding tightly onto coral with its tail.

Fan coral

A **conch** hides in the seagrass.

Blue ribbon eel

Sand is tiny, crumbled pieces of shell or rock. Not all sand is yellow. Some beaches have pink, white, black, or even green sand.

Wow!
FACT

The **Great Barrier Reef** off northeast Australia is the world's largest coral reef. It is more than 1,400 miles (2,300 km) long—that's about half the size of the state of Texas. Astronauts can see it from outer space.

Amphibians

The word "amphibian" comes from a Greek word meaning "both lives." Amphibians live part of their lives in the water and part on land. Almost all amphibians are born under water in ponds, lakes, or streams. When they are born, they are tadpoles. When they grow older, they leave the water and live on land. There are three types of amphibians—**frogs** and **toads, salamanders,** and **caecilians**.

AMPHIBIANS:
- are vertebrates
- are cold-blooded
- breathe with gills, lungs, or through their skin
- go through metamorphosis
- lay eggs

Most amphibians spend their whole lives near freshwater and damp areas. Even if they go into in fields and woodlands to hunt for insects to eat, they never wander too far from water. Their thin skin must stay moist in order for them to breathe.

Ology Alert!

Herpetology is the study of amphibians and reptiles.

From Tadpole to Frog

The process from baby to adult is called **metamorphosis**. All amphibians go through metamorphosis.

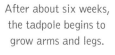

Frogs and toads make up almost 90% of all amphibians.

The female frog mates with the male frog. Then she lays her eggs in the water. Amphibian eggs have no shells, so the water keeps them moist.

The eggs hatch and tiny tadpoles come out. Tadpoles have no arms, legs, or lungs. They breathe through gills and live entirely underwater. They have a tail and fins.

After about six weeks, the tadpole begins to grow arms and legs.

At about twelve weeks, its tail shrinks and it develops lungs.

It is now a frog. The frog moves onto land and now breathes through lungs.

Wow!
F A C T

The **spadefoot toad** smells like peanut butter.

What's the Difference?

Wow!
F A C T

Frogs and toads usually shed their skin about once a week— and then eat it!

FROG	TOAD
Moist smooth skin	Dry, bumpy skin
Thin shape	Fat shape
Large hind legs, so they can jump far	Short hind legs. Rarely jumps and when it does, takes short hops
Leaps away from predators	Stays motionless when spots predators
Lives in and around water	Lives away from water

A frog captures insects and spiders with its long, sticky tongue. It swallows its prey whole.

Only male frogs and toads croak. They croak to attract females. Each species makes a different sound.

The call of the male **Natterjack toad** can be heard many miles away.

Tiger salamander

Frog Questions Answered

Do frogs cause warts?
No. People probably thought this because frogs have bumpy, warty-looking skin. Warts are caused by a virus.

Poison dart frog

What happens if I kiss a frog?
You will get a gross taste on your lips, but no prince will appear. If you kiss a **poison dart frog**, though, you could die.

Are brightly-colored frogs poisonous?
Almost always!

Why do frogs pee on you when they're picked up?
They don't want to be picked up. By peeing, they are trying to get your hands off them so they can hop away.

Salamanders

The salamander may look like a lizard, but it's an amphibian. Unlike lizards, salamanders do not have claws and their skin is moist and smooth. They also lay their eggs in the water. All salamanders have a tail that they can regenerate, or grow back if it is torn off by a predator.

A **newt** is a kind of salamander. It has a flatter tail and bumpier skin than other salamanders.

HALL OF FAME

Largest amphibian: The **Chinese giant salamander** can grow to 6 feet (2 m) long.

Smallest amphibian: The **Paedophryne amanuensis** is also the world's smallest vertebrate animal. It is about 1/3 inch (8 mm) long.

Deadliest frog: The **golden poison frog** has enough poison to kill 10 to 20 people. If a person touches this frog without gloves, he or she could die.

Biggest frog: The **goliath frog** of West Africa can grow to more than 1 foot (30 cm) in length and weigh more than 7 pounds (3 kg). It also has the longest leap, jumping 10 feet (3 m) in a single hop.

Caecilians

Caecilians look like worms, but they are also amphibians. They live in tropical areas in underground tunnels that they dig with their pointed heads.

Freshwater

The freshwater biome includes lakes, ponds, rivers, and streams. Water is called "fresh," if it contains less than 1% salt. Freshwater comes from rain or melted snow. Freshwater is either still (not moving) or moving.

A **pond** is still water. It is shallow, so sunlight can reach the bottom, allowing underwater plants to grow.

A **lake** is larger and deeper than a pond. Most lakes have still water, but larger lakes have some moving water.

The **egret** has long, slim legs for wading in shallow water. It has a long neck and sharp beak to stab and grab fish and frogs.

Cattails have roots that anchor them in the mud near the shoreline.

The **wood duck** lays its eggs in a hole in a tree that is near or even hanging over a pond. When the eggs hatch, their mom calls to them from the pond. The ducklings then leap from as high as 50 feet (15 m) to meet her in the pond below.

Trout are strong swimmers. Their tails help them glide through a river's rushing waters. Trout have trouble getting the oxygen from the water into their gills, because they can't move their fins. Instead, they must swim quickly to push oxygen into their gills.

River otters like to play games in the water.

Northern water snakes are often covered in mud. They are not poisonous and like to hide near beaver lodges.

The **pond skater** is an insect that can walk on water! It uses its middle legs like paddles.

Rivers and streams are always moving. A river starts up high on a mountain or a hill. Snow collects on the mountain and melts. The melted snow becomes a stream. As the stream flows downhill, it picks up speed. It gains water from other streams and from rainfall that has collected on the ground. When one stream meets another, they join together. The smaller stream is now known as a **tributary**. Many tributary streams are needed to form a river. Most rivers eventually flow into an ocean, sea, or large lake. The end of the river, where it meets a lake or an ocean, is called the **mouth**.

Algae Doctor

Have you ever seen a pond with yellowish-green water? That pond has *a lot* of algae in it. If you place a drop of pond or lake water under a microscope, you will see tiny algae. Algae don't have leaves, stems, or roots like a plant, but they make food using energy from the Sun like plants do. This energy is passed onto freshwater animals when they eat the algae. Algae also make oxygen that helps the underwater animals breathe. You can tell if a pond is healthy by the amount of algae living in it—not too little and not too much.

Spotlight: Beaver

Have you ever heard the phrase "busy as a beaver"? Well, the beaver is a very busy animal. The beaver works all night. First, it chops down trees with its long buck teeth. Then it uses the trunks to build a dam in a river or stream. The dam creates a pond of still water, where the beaver builds its home, called a lodge. The dome-shaped lodge is made of twigs and mud. Its entrance is underwater to keep predators out. Most lodges have two rooms—one for drying off after entering and one where the beaver family lives. The beaver uses its paddle-shaped tail like a rudder on a boat to steer itself as it moves logs along the river. A beaver can stay underwater for 15 minutes, because it has valves in its ears and nose that close when underwater. Plus the beaver has built-in swim goggles—a clear layer that covers its eyes and allows it to see underwater.

Water Pistol

The **archerfish** shoots a stream of water out of its mouth to knock an insect off a branch above. When the insect topples into the water, the archerfish eats it. The archerfish can hit a target up to 10 feet (3 m) away.

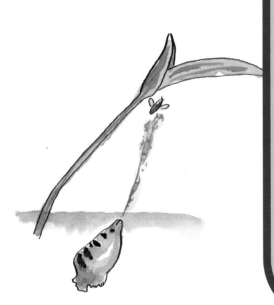

53

Reptiles

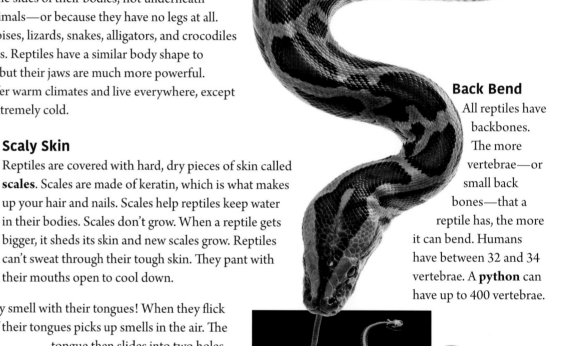

Tiger python

The word "reptile" comes from a Latin word meaning "to crawl." Reptiles crawl or slither with their bellies on the ground, because their short legs are attached to the sides of their bodies, not underneath like other animals—or because they have no legs at all. Turtles, tortoises, lizards, snakes, alligators, and crocodiles are all reptiles. Reptiles have a similar body shape to amphibians, but their jaws are much more powerful. Reptiles prefer warm climates and live everywhere, except where it is extremely cold.

REPTILES:
• are vertebrates
• are cold-blooded
• have scaly skin
• breathe through lungs
• lay soft, leathery eggs, but some give birth to live young

Scaly Skin

Reptiles are covered with hard, dry pieces of skin called **scales**. Scales are made of keratin, which is what makes up your hair and nails. Scales help reptiles keep water in their bodies. Scales don't grow. When a reptile gets bigger, it sheds its skin and new scales grow. Reptiles can't sweat through their tough skin. They pant with their mouths open to cool down.

Back Bend

All reptiles have backbones. The more vertebrae—or small back bones—that a reptile has, the more it can bend. Humans have between 32 and 34 vertebrae. A **python** can have up to 400 vertebrae.

The Nose-tongue

Many reptiles have a forked tongue that they use for smelling. That's right—they smell with their tongues! When they flick their tongues out of their mouths, the tip of their tongues picks up smells in the air. The tongue then slides into two holes, called Jacobson's organ, in the roof of the mouth. This organ allows reptiles to figure out the smell. Smells help them find food.

Brain

Jacobson's organ

Tongue

Python skeleton

Coral snake

Sun and Shade

Because reptiles are cold-blooded, their body temperature is the same as the temperature around them. Reptiles sit in the sun to warm up and go in the shade or cold water to cool down.

Turtles and Tortoises

Turtles and tortoises haven't changed much since they first appeared on earth more than 200 million years ago. Their survival is probably due to their shell, which acts like a shield. A turtle's shell is made up of 60 different bones all connected together. The shell also has nerve endings. If you touch the shell, a turtle can feel it. Many turtles hide inside their shells when attacked by predators.

Wow!
F A C T

The **North American box turtle** can live to be over 100 years old.

What's the Difference?

TURTLE	TORTOISE
Lives in freshwater, saltwater, or land—but most turtles live in freshwater.	Only lives on land. They cannot swim.
Omnivore	Herbivore
Flatter shell	Rounder shell

Snakes

Snakes are long, skinny reptiles with no arms or legs. They move by flexing and scooting their body. Snakes also have no eyelids. Instead, they have transparent skin that protects their eyes. They never blink, and they sleep with their eyes open.

Many snakes are constrictors. A **boa constrictor** coils its body around and around an animal and squeezes the life out of it. Then it swallows it whole. A snake's jaw unhinges, so it can open its mouth very wide to swallow big prey. If it has eaten a large animal, a snake can go weeks or even months without eating again.

Some snakes are poisonous. Most poisonous snakes are brightly colored to warn away predators. A wide, triangular head usually means the snake is poisonous—but this clue doesn't always work. **Coral snakes** and **cobras** are poisonous, and they have oval heads.

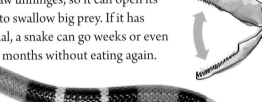

Rattlesnakes are poisonous. They have a rattle made of hard scales at the end of their tails. The rattlesnake shakes its rattle to warn away predators. It grows a new rattle every time it sheds its skin.

Rattle

Rattlesnake

Copperheads belong to a poisonous group of snakes called **pit vipers**. They have heat-sensing pits behind their nostrils to help them hunt.

Lizards and Iguanas

Most lizards have four legs, eyelids, a square or triangular head, and a long tail. A lizard uses it tail to help balance, run, and swim. If a lizard's tail breaks off in a fight, the tail will keep wiggling and confuse the enemy, while the lizard runs to safety. Later, the lizard will grow a new tail. Lizards eat insects and rodents.

Some lizards, such the **green crested lizard**, have a frilly crest on their backs.

Chameleon

A **chameleon** can change color to match its surroundings. It doesn't do it to hide from predators. It changes color depending on the air temperature, the amount of light, and its mood. Also, a chameleon's eyes can point in two different directions at one time, so it can better search for food and escape other animals.

Gecko

A **gecko** has sticky pads with thousands of tiny, stiff hairs on its toes. The hairs stick to most surfaces, so it can walk upside-down on the underside of a branch or on ceilings.

The **Komodo dragon** is the largest lizard in the world. It eats pigs, deer, and even water buffalo. It can swallow a piece of food bigger than its head!

Komodo dragon

The **Gila monster** and the **Mexican beaded lizard** are the only venomous lizards.

Mexican beaded lizard

Crocodiles and Alligators

Both alligators and crocodiles are fierce hunters. They drown their prey by spinning them around in the water. Their bite is so strong that they can crush a metal canoe. It's easy to confuse an alligator and a crocodile, since these two reptiles look a lot alike.

What's the Difference?

ALLIGATOR	CROCODILE
Brown	Black
U-shaped snout	V-shaped snout
Large, pointy bottom tooth is hidden when mouth is closed	Large, pointy bottom tooth is seen when mouth is closed
Lives in freshwater	Lives in saltwater and freshwater

Wow!
FACT

Both alligators and crocodiles store fat in their tails. They can live off this fat for months if they can't find an animal to eat.

Get to Know: Steve Irwin
(1962–2006)

Wildlife expert **Steve Irwin** was called "The Crocodile Hunter." Steve grew up in the reptile park in Australia that his parents owned. As a child, one of his chores was to feed the reptiles. And guess what his parents gave him for his sixth birthday present? A large python! By age nine, Steve was helping his dad wrestle crocodiles that swam too close to the boat ramps. In his twenties, he traveled around Australia, trapping crocodiles living in areas with people and bringing them to the reptile park, so they wouldn't be killed for their skin. He took over running the park and renamed it Australia Zoo. One day, Australian TV showed a video of him catching a huge crocodile. People were amazed. They wanted to know more about The Crocodile Hunter. Soon, Steve had his own successful TV show that aired all over the world. He wanted to make people passionate about reptiles and animals, so they would help to keep them safe. He was killed by a stingray in 2006, while filming a movie about deadly ocean creatures.

Deserts are some of the driest spots on Earth. Most deserts get less than 10 inches (25 cm) of rain a year. The Atacama Desert in Chile is the driest place of all. In some parts of the Atacama, there was no rain for 400 years! When rain does fall in the desert, it often **evaporates**, or disappears as water vapor, before it reaches the ground. As a result, there is little plant and animal life.

Deserts

Hot and Cold

Many deserts are sizzling hot. How hot? In some deserts, you can fry an egg on a rock! The Sahara Desert recorded temperatures of 136.4°F (58°C). Many hot deserts turn very cold at night, because there is no humidity. And some deserts are cold all the time.

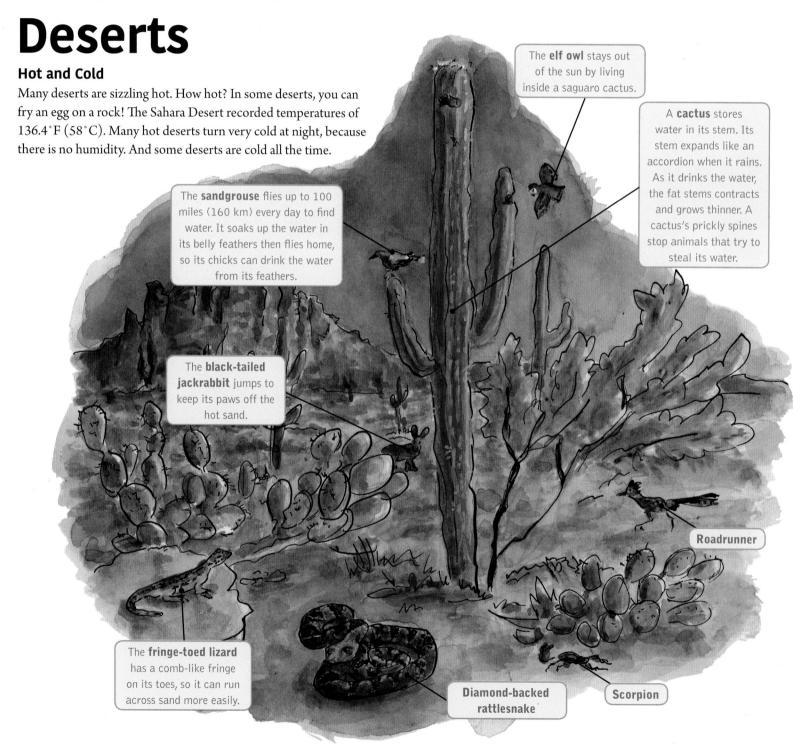

The **elf owl** stays out of the sun by living inside a saguaro cactus.

A **cactus** stores water in its stem. Its stem expands like an accordion when it rains. As it drinks the water, the fat stems contracts and grows thinner. A cactus's prickly spines stop animals that try to steal its water.

The **sandgrouse** flies up to 100 miles (160 km) every day to find water. It soaks up the water in its belly feathers then flies home, so its chicks can drink the water from its feathers.

The **black-tailed jackrabbit** jumps to keep its paws off the hot sand.

The **fringe-toed lizard** has a comb-like fringe on its toes, so it can run across sand more easily.

Roadrunner

Diamond-backed rattlesnake

Scorpion

Where's the Water?

If all living things need water to survive, what happens in the dry desert? Hardy desert plants and animals find ways to search out and save precious water.

The **baobab tree** can store water in its trunk for nine months.

The **dung beetle** gets moisture from eating other animals' poop.

It's Tough Out There

Survival isn't easy in the desert. Not only is it tricky to find water and stay cool, it's just as tricky to hunt for food and avoid becoming another animal's meal.

The **scorpion** has a stinger at the end of its tail that releases venom. This venom can kill insects instantly.

Roadrunners are super-speedy. They can outrun and capture lizards, fast-moving snakes and rodents, and small birds.

Sandy . . . or Not

Tons of sand covers many deserts. The strong desert winds wear down rock, creating tiny pebbles. Over time, the pebbles crumble into sand. Wind blows the sand into giant heaps, called **sand dunes**. But not all desserts are sandy. Some deserts have rocks or gravel. Some have grasses. And some even have snow! Antarctica is considered a desert by many scientists. Why? The snow here rarely melts, because the air is so dry.

Spotlight: Camels

Camels are one of the few large mammals that live in the desert. They are sometimes called "the ships of the desert," because humans use them to carry heavy loads. Their wide, padded feet do not sink into the sand. They have two rows of long eyelashes and special eyelids that keep sand from blowing into their eyes. They can close their slit-like nostrils so sand doesn't blow up their noses.

Camels can go up to 20 days without water. (Humans can only go three to five days.) Camels do not store water in the humps on their backs. The humps are filled with fat. The fat gives them energy in case they can't find food. Camels can smell water from miles away, and they can drink 30 gallons (114 l) in ten minutes! That is the same as you drinking 480 glasses of water.

Arabian camels (from northern Africa) have one hump.

Bactrian camels (from Asia) have two humps.

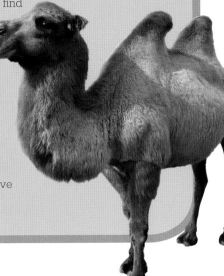

Beat the Heat

Most desert animals disappear into burrows underground or hide beneath rocks to escape the daytime sun. Desert animals usually have a sandy color. The light color helps reflect sunlight and heat.

The **fennec fox** uses its big ears to keep cool. Heat flows out of its body through its ears.

Wow!
F A C T
Deserts are often extremely windy. In 2006, a sandstorm in the Gobi Desert carried 330,600 tons of sand!

Birds

BIRDS:
- are vertebrates
- are warm-blooded
- have two wings, two legs, and a beak or bill
- lay eggs
- can fly (except for a few)

Look up, down, and all around... birds are everywhere. They're in the sky, on telephone wires, paddling in ponds, running across sand, standing on ice, singing in forest trees, and in your own yard.

Wow!
F A C T
The **black woodpecker** strikes its bill against a tree between 8,000 and 12,000 times a day.

Feet and Toes

Most birds have four toes—three that face forward and one that points backward—so they can grasp onto a branch or twig.

Waterfowl, or birds that live in the water, have webbed feet to paddle better.

Carnivorous birds have sharp **talons**, or claws, that they use to puncture or hold down their prey.

Beaks and Bills

Birds do not have teeth or jaws. Instead, they have bills or beaks. The shape of a bird's bill or beak depends on the food that it eats.

- A **red-tailed hawk** has a sharp, hooked beak for tearing meat.

- A **finch** has a cone-shaped bill to crack seeds.

- A **woodpecker** has a long, skinny bill to poke holes in wood to look for insects.

- A **hummingbird** has a long, straw-like beak to suck nectar from flowers

- A **duck** has a flat bill to eat plants in the water.

A **pelican** stores fish in the big pouch in its beak.

A nest usually has many layers:

Coarse twigs are on the bottom.

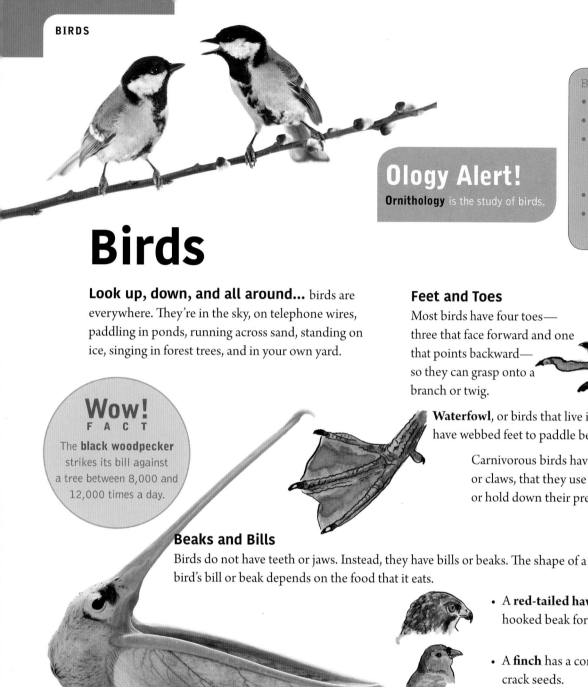

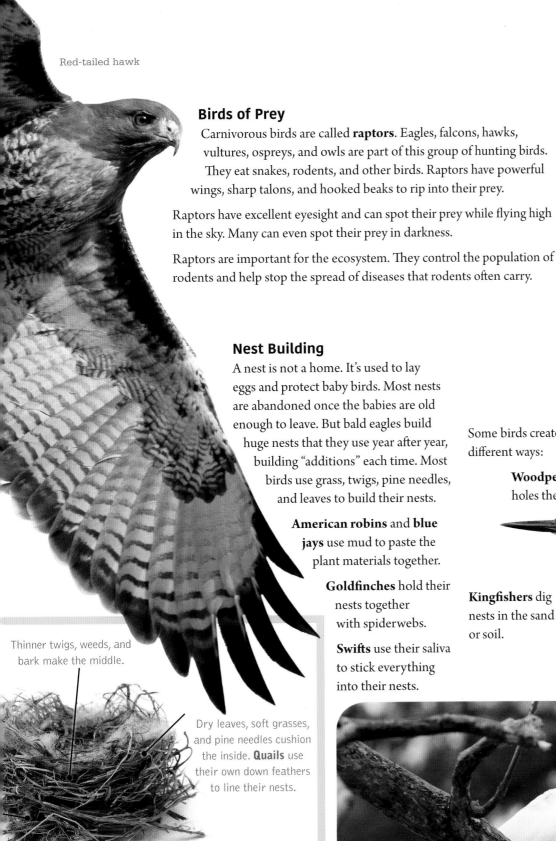

Red-tailed hawk

Birds of Prey

Carnivorous birds are called **raptors**. Eagles, falcons, hawks, vultures, ospreys, and owls are part of this group of hunting birds. They eat snakes, rodents, and other birds. Raptors have powerful wings, sharp talons, and hooked beaks to rip into their prey.

Raptors have excellent eyesight and can spot their prey while flying high in the sky. Many can even spot their prey in darkness.

Raptors are important for the ecosystem. They control the population of rodents and help stop the spread of diseases that rodents often carry.

Nest Building

A nest is not a home. It's used to lay eggs and protect baby birds. Most nests are abandoned once the babies are old enough to leave. But bald eagles build huge nests that they use year after year, building "additions" each time. Most birds use grass, twigs, pine needles, and leaves to build their nests.

American robins and **blue jays** use mud to paste the plant materials together.

Goldfinches hold their nests together with spiderwebs.

Swifts use their saliva to stick everything into their nests.

Thinner twigs, weeds, and bark make the middle.

Dry leaves, soft grasses, and pine needles cushion the inside. **Quails** use their own down feathers to line their nests.

Angry Birds

What if a bird doesn't feel like making a nest? **Cuckoos** are the bullies of the bird world. A cuckoos lays its eggs in another bird's nest—even while the other bird is sitting there! Then the cuckoo leaves, forcing the other bird to raise her baby. If that bird tries to get rid of the cuckoo's egg, the cuckoo flies back and hurts the bird or her eggs! And if the cuckoo egg hatches first, the baby cuckoo will often throw the other eggs out of the nest to get all the attention of its adopted parents.

Some birds create nests in different ways:

Woodpeckers live in the holes they make in trees.

Kingfishers dig nests in the sand or soil.

The **white tern** sticks its eggs between two tree branches.

Up, Up, and Away . . .

All birds have feathers and wings, yet not all birds can fly. Flight gives birds an advantage over other animals. They can fly to find food and water and get away from bad weather. Baby birds know how to fly without being taught. But they do need to practice steering and landing!

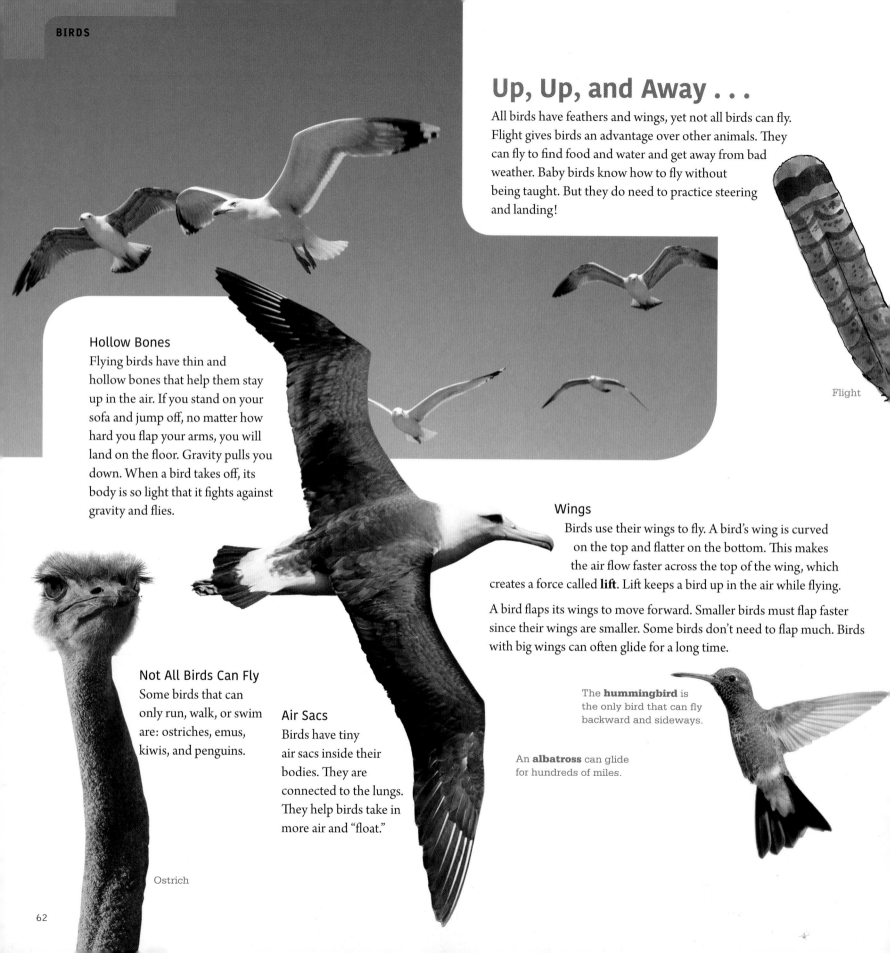

Flight

Hollow Bones

Flying birds have thin and hollow bones that help them stay up in the air. If you stand on your sofa and jump off, no matter how hard you flap your arms, you will land on the floor. Gravity pulls you down. When a bird takes off, its body is so light that it fights against gravity and flies.

Wings

Birds use their wings to fly. A bird's wing is curved on the top and flatter on the bottom. This makes the air flow faster across the top of the wing, which creates a force called **lift**. Lift keeps a bird up in the air while flying.

A bird flaps its wings to move forward. Smaller birds must flap faster since their wings are smaller. Some birds don't need to flap much. Birds with big wings can often glide for a long time.

Not All Birds Can Fly

Some birds that can only run, walk, or swim are: ostriches, emus, kiwis, and penguins.

Air Sacs

Birds have tiny air sacs inside their bodies. They are connected to the lungs. They help birds take in more air and "float."

The **hummingbird** is the only bird that can fly backward and sideways.

An **albatross** can glide for hundreds of miles.

Ostrich

62

Tail

Feathers

Feathers help a bird fly. Tail feathers are used for steering, just like a rudder in a sailboat. Long feathers, called flight feathers, make the wingspan bigger. Semiplume feathers keep a bird warm and give it shape. Bristle feathers are usually found on a bird's head. Short, down feathers keep a bird warm. Once a year, a bird's feathers **molt**, or fall out—a few at a time— and new feathers grow in.

Semiplume

Get to Know: John James Audubon
(1785–1851)

John James Audubon was an ornithologist who was famous for his paintings of birds. He published a book of 435 life-sized, realistic watercolors of North American birds, titled *Birds of America*. Audubon was the first person in North America to band birds. He tied silver thread to the legs of Eastern phoebes. This way he could keep track of the birds. He learned that birds return to the same nesting place each year. After he died, the Audubon Society was named in his honor. The Audubon Society helps to protect birds all over the world.

Takeoff

The hardest part of flying is taking off. Birds try to take off facing the wind. The wind gives them an extra push, so they don't need to flap their wings as hard. Lighter birds have an easier time taking off than heavier birds.

A **goose** has to paddle across the water.

A **sandhill crane** runs along the ground to gain enough speed.

A **puffin** stands on the edge of a high cliff to catch a gust of wind.

HALL OF FAME

Biggest flying bird: The **California condor** can fly 150 miles (250 km) in one day.

Smallest bird: The **bee hummingbird** is only 2.5 inches (6 cm) long. It's eggs are smaller than your fingernail.

Fastest flying bird: The **peregrine falcon** can fly 200 miles per hour (320 kph). That's the same speed as the fastest train.

The highest flyer: A **Rüppell's griffon vulture** once crashed into an airplane!

Largest bird: **Ostrich** can grow up to 8 feet (2.5 m) tall. That's taller than the tallest human.

Largest egg: The ostrich lays an egg that weighs about 3 pounds (1.5 kg)—the same as a medium-sized cantaloupe!

Forests

Forests are large, tree-covered areas.

One-third of all land is forests. There are three kinds of forests—coniferous forest, deciduous forest, and tropical rain forest.

Coniferous Forest

What kind of trees are in a coniferous forest? **Conifers**, of course! Conifers have needles instead of leaves and grow cones instead of flowers. Pine, spruce, and fir trees are conifers. Conifers are decorated for Christmas.

Conifers are found where there are long, cold winters. Most conifers are **evergreen**, which means they keep their needles all year long. Their needles have a waxy coating that protects them during the cold winters. Tall, narrow conifers grow close together to stay safe from icy winds.

Other names for the coniferous forest are "taiga" and "boreal forest."

The **crossbill** uses its crossed beak to pluck seeds from pinecones.

The **moose** is the largest member of the deer family. Every winter, the male sheds its heavy antlers and grows new ones in the spring.

Deciduous Forest

Deciduous trees have broad, flat leaves that fall off when the weather grows cold and grow back when the weather warms up. Deciduous leaves also change color with the seasons. Why do they do this?

In the cold autumn and winter, deciduous trees can't get enough water or light, so the sap inside the trunk stops flowing to the leaves. Without the sap, green chlorophyll in the leaves disappears and bright colors come out. Then the leaves fall off. The tree becomes **dormant**, or stops growing. In the spring, the sap starts flowing from where it had been stored in the roots, and the tree grows again. Deciduous trees grow best in climates where there are four seasons.

Red maple and scarlet oak—red

You can tell the type of tree by the color its leaves turn in autumn:

Poplar and birch—yellow

Sugar maple— orange

Willow— stays green

Dogwood— maroon

Tropical Rainforest

Tropical rainforests are thick, warm, and wet forests. Most tropical rainforests are found near the equator (an imaginary line that circles the Earth), and they have a hot, muggy climate.

Rainforests contain more than half of the world's plant and animal species!

1 Emergent layer

2 Canopy

3 Understory

4 Forest floor

There are four layers to a rainforest:

1 **The emergent layer.** Only the tallest trees reach this very sunny top layer.
Home to: birds, butterflies, bats, small monkeys

2 **The canopy.** Raindrops often get trapped in the thick branches and leaves. It can take ten minutes for a raindrop to fall from the canopy to the forest floor.
Home to: birds, monkeys, sloths, snakes, lizards

3 **The understory.** A shady place that is filled with vines, small trees, and the trunks of taller trees.
Home to: tree frogs, snakes, insects

Red-eyed tree frog

Leaf-cutter ants

The **pitcher plant** eats insects and small animals, such as frogs. Sweet nectar in its pitcher-shaped pot attracts the animals. Once inside, they slip into the pot and drown in the juices in the bottom. The plant digests their bodies.

Harpy eagle

The **blue-and-yellow macaw** is part of the parrot family. Its loud calls can be heard throughout the rainforest canopy.

The seeds of the **cocoa fruit** are used to make chocolate.

Three-toed sloth

The **marmoset** is a mischievous, playful monkey.

The **tapir** looks like a pig, but it's a relative of the horse. It picks fruit and leaves with its short trunk.

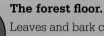

Jaguar

4 **The forest floor.** Leaves and bark cover the ground. Barely any sunlight reaches down here, so it is damp and dark.
Home to: mushrooms and earthworms, jaguars, gorillas, tapirs, and tarsiers.

MAMMALS:
- are vertebrates
- are warm-blooded
- have fur or hair
- give birth to live young (except for two species)

Ology Alert!
Zoology is the study of animals.

Mammals

A tiny mouse and a huge rhinoceros—what do they have in common? They're both mammals. Mammals may look extremely different from one another, but they all share these traits:

- All mammals have hair on their bodies. Fur, whiskers, spines, and horns all count as hair. A fur coat helps keep a mammal warm by slowing down the heat escaping from its body.

- Most mammals give birth to live young. All mammal babies drink milk that comes from their mothers' bodies.

- All mammals use lungs to breathe. Mammals that live in water must come up to the surface to breathe air.

- Most mammals have large brains which makes them more intelligent than most other animals.

There are three groups of mammals: **monotremes, marsupials,** and **placenta** mammals. They are grouped by how they give birth and take care of their young.

Echidna

Group of Two
The **platypus** and the **echidna** (also called **spiny anteater**) are **monotremes**. They are the only mammals that lay eggs. Monotremes are the most primitive mammals and are closely related to reptiles. Today, the platypus and the echidna are only found in Australia and New Guinea.

In My Pocket
The kangaroo, koala, opossum, and the Tasmanian devil are all **marsupials**. Marsupial mammals have a pouch, or a pocket. When a marsupial baby is born, it climbs into its mother's pouch and spends several months inside, nursing on its mother's milk and growing strong enough to survive on its own.

The **kangaroo** hops with powerful back legs. Large kangaroos can jump over a 10-foot (3-m) fence. They can hop faster than a galloping horse. Male kangaroos box when they fight. They stand on their hind legs and punch each other. The kangaroo that is left standing is the winner.

Some people call the **koala** a "koala bear," but it is not really a bear. A koala sleeps for up to 20 hours a day in the branches of a eucalyptus tree. Its special digestive system lets it eat poisonous eucalyptus leaves.

Koalas

Giant Siberian tiger

Live Young

Most mammals are **placental mammals**. Placental mammals give birth to live young. You are a placental mammal.

The Feline Family

Felines come in all sizes, from the small house cat to the **giant Siberian tiger**. A cat's whiskers can sense movement in the air. The whiskers help it move around in the dark. Cats have great hearing. They can move each ear separately in the direction of the sound.

Cats are excellent hunters. They have **retractable claws**. They can pull their sharp claws into their paws when they are not using them, so they don't get worn down. The **cheetah** is the only cat that doesn't retract its claws.

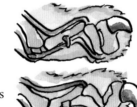

Cat

Bats

Bats are the only mammals that can truly fly. The flying squirrel may look as if it flies, but it only glides from branch to branch. Bats hang upside-down. Why? One reason may be that their leg bones are too thin to support their bodies.

Fruit bat

Elephants

There are two kinds of elephants—African and Asian. The African elephant is larger and mostly lives on the savanna. The Asian elephant mostly lives in forests.

An elephant's trunk is a combination of an upper lip and nose. There are over 40,000 muscles in an elephant's trunk. An elephant uses its trunk for smelling, picking up food, plucking fruit off trees, drinking water, and greeting other elephants. On a hot day, an elephant can suck up water with its trunk and give itself a shower!

The Canine Family

Do you have a dog? Your dog is part of the canine family, along with coyotes, wolves, jackals, and foxes. Canines have long, pointed snouts and padded feet with non-retractable claws. Unlike humans and bears that walk on the soles of their feet, canines walk on their toes.

All dogs have a great sense of smell. A dog's nose is thousands of times more sensitive than a human's.

Golden retriever puppy

African elephant

Rodents

Mice, rats, hamsters, chipmunks, squirrels, and beavers are all rodents. Rodents usually have long tails and have two strong front teeth that never stop growing. Rodents gnaw on wood and other tough surfaces to stop their teeth from becoming too long. There are more rodents than any other kind of mammal. One mouse can give birth to 100 babies in only one year!

What's the Difference?

MOUSE	RAT
Smaller	Larger
Pointy snout with long whiskers	Rounder snout
Long, skinny, hairy tail	Long, thicker, hairless tail
Large ears compared with head size	Small ears compared with head size
Small poop	Large poop

A **hamster** stores nuts and berries in pouches in both cheeks to bring home to eat later. The food stays dry because hamsters don't have saliva in their cheeks. **Chipmunks** and **ground squirrels** also have cheek pouches. A chipmunk's pouches can puff out to three times the size of the chipmunk's head when they are full.

Rabbits and hares are not rodents. Why? Mostly because they have four large teeth, instead of two.

The **capybara**, which lives in South America, is the largest rodent and can weigh up to 145 pounds (66 kg).

Primates

Monkeys and apes are primates. The biggest difference between the two is that apes don't have tails. Also, apes can swing from branch to branch in trees because they have shoulders similar to human shoulders. Most monkeys climb or run on the branches. Apes tend to be larger and smarter, and their skeletons look similar to human skeletons. Monkey skeletons look more like dog or cat skeletons. All primates have thumbs or big toes that stick out to the side to let them grasp things, the same way humans do.

Orangutan   **Chimp**   **Gorilla**   **Human**

Monkeys

Many monkeys have a **prehensile tail**, which means they can grasp things with it. They use their flexible tail like a fifth hand to hold onto branches and vines, as they climb high into the treetops.

Scientists don't know why the **proboscis monkey** has such a large nose.

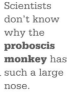

A **spider monkey** has an extremely long tail and long arms and legs.

The colorful **mandrill** is very shy.

Apes

There are four types of great apes:

Gorillas are the largest primate, and, for the most part, they are gentle. They eat leaves, fruit, and bamboo. A gorilla's arms are longer than its legs, so it uses its knuckles on its hands to walk on all fours. An adult male gorilla is called a silverback, because of silvery hair on its back. The silverback is the leader of its family, and it scares away other animals by standing and beating its chest.

Gorilla

Chimpanzees are the smartest land mammals, after humans. They use simple tools to solve problems. For example, a chimp will use a long, thin stick to poke termites out of their mounds. It will use a rock to crack nuts and squeeze leaves into a ball to soak up water to drink. Chimpanzees are very social and live in large groups. They communicate with sounds and facial expressions.

"Orangutan" means "person of the forest" in the Malay language of Indonesia. Many people think orangutans look like hairy old men.

Bonobos look a lot like chimpanzees. There are very few alive today, and they are only found in the Democratic Republic of Congo. Bonobos are sweet and cooperative.

Get to Know: Jane Goodall
(1934–present)

Jane Goodall is a famous primatologist, a scientist who studies primates. She studied chimpanzees in Tanzania, Africa. As a child, Jane dreamed of visiting Africa. When she was 23 years old, her dream came true when a friend invited her to Kenya. Jane got a job helping archaeologist Louis Leakey. She moved to the Gombe Stream National Park in Tanzania to study wild chimpanzees. She lived at a campsite. Every day, dressed in all-tan clothes to blend in, she went into the forest to look for the chimps. She sat quietly for hours, but the chimps always stayed far away. If she tried to move close, they'd run away. But she sensed they were watching her, too. Months and months went by. Jane could now tell the chimps apart. She gave them names. She understood their grunts and screams. Finally one day, a chimp came up to her. She held out a piece of fruit. He took it, and then he held her hand. This was the beginning of a long friendship. Jane spent years studying the chimps in great detail. She discovered that chimps are omnivores, which means they eat fruits, nuts, leaves, insects, and small animals. She also discovered that chimps make and use tools. Up until then, scientists thought only humans made tools. Jane wrote many books and educated the world about chimpanzees.

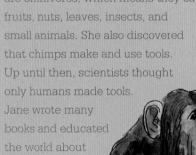

Chimpanzees

69

Hippopotamus

Hoofed Mammals

A hoof is basically a huge toenail. Unlike a claw or a nail, the hoof fully touches the ground and supports all the weight of the leg. Hooves allow an animal to run faster. Hoofed mammals, also known as **ungulates**, are some of the largest mammals. They are divided into two groups: even-toed and odd-toed.

Odd-toed Mammals Stand on One or Three Toes

Horses are domesticated animals, meaning they live with humans. Wild horses, also called mustangs, live in herds on grasslands. Horses have bigger eyes than most mammals. Because their eyes are on the sides of their heads, they can see in almost a full circle.

A **zebra** is part of the horse family. Every zebra has a different pattern of stripes. As soon as a zebra is born, it learns its mother's pattern so it can find her in a group.

Horse

Even-toed	Odd-toed
antelope, cattle, camel, deer, giraffe, goat, hippopotamus, hog, llama, pig, sheep	donkey, horse, mule, rhinoceros, tapir, zebra

Even-toed Mammals Stand on Two or Four Toes

The word "**hippopotamus**" comes from two ancient Greek words that mean "river horse." These plant-eating giants spend much of their lives in the shallow water of slow-moving rivers. When a male hippo opens its mouth wide and shows it teeth, it is a warning to stay away. The hippo has its eyes, ears, and nostrils on top of its head, so it can stay mostly underwater and still know what's going on.

Pig

A **pig** is a very clean animal. It rolls in the mud to cool off. A pig uses its snout to smell and to dig for insects, roots, and worms.

Zebras

Mammals—Not Fish!

Some mammals live in the water. Because mammals have lungs, they need to swim up to the surface to breathe. Whales and dolphins breathe using blowholes at the top of their heads. Water mammals have fins or flippers and are excellent swimmers.

There are two types of whales: toothed and baleen. **Beluga whales** have teeth. **Humpback whales** do not have teeth. Instead they have baleen. Baleen acts like a coffee filter, straining out food from the water. The humpback whale eats several million shrimp-like organisms called krill at a time.

Dolphins are sometimes called the "dogs of the sea," because they are so playful. Dolphins are extremely smart, and "talk" with other dolphins using clicking and whistling sounds.

A **manatee** can stay underwater for up to 20 minutes before it must swim to the surface to breathe. These big, slow mammals are sometimes called "sea cows." Some manatees live in freshwater and some live in saltwater, but all eat plants.

Dolphin

Beluga whale

Manatee

Cheetah

HALL OF FAME

Largest land mammal: The male **African elephant** stands up to 13 feet (4 m) tall and weighs up to 13,000 pounds (5,900 kg). It weighs the same as a Tyrannosaurus rex did!

Largest marine mammal: The **blue whale** measures up to 100 feet (30 m) long. Its heart is the size of a small car!

Smallest mammal: The **bumblebee bat** is the size of a bee, about 1 inch (2.5 cm) long.

Fastest digger: The **badger**, with its sharp claws and partly webbed front toes, can dig an underground tunnel faster than a human can with a shovel.

Fastest land mammal: The **cheetah** can go from 0 to 60 miles per hour (96 kph) in only 3 seconds.

Fastest marine mammal: The **orca**, part of the dolphin family, can swim 30 miles per hour (48 kph).

Slowest mammal: The **sloth** moves so slowly that algae grows on its fur.

Grasslands

Grassland are exactly what they sound like— open, flat areas covered in a carpet of tall grass with few trees. Grasslands are often found in an area that is too dry to be a forest and not dry enough to be a desert. Grasslands are very windy.

Bottoms Up!

Grass grows differently from many other plants. Most plants grow from their top—from their stems or leaves— but grass grows from its bottom. So if an animal munches off the top of the grass, new growth doesn't stop.

Down Below

Because there a few good hiding places in the open grasslands, the **prairie dog** lives underground in burrows.

Prairie dog

The grasslands biome is divided into **temperate grasslands** (prairies, pampas, steppes) and **tropical grasslands** (savannas).

What's the Difference?

TEMPERATE GRASSLANDS	TROPICAL GRASSLANDS
No trees, because there's little moisture in the soil.	Many scattered single trees and shrubs
Hot summers and cold winters	Hot all year long with a rainy season

A Big Salad Bar

Many animals can live together in grasslands because they each eat a different part of the plant. Some eat up high and some down low.

Safety in Numbers

Imagine you are standing alone in the middle of the empty savanna. A predator spots you. It's just him and you. You'd better be a fast runner! Now imagine that you aren't alone. It's the predator against you and 20 of your friends. Unless you are the slowest runner of your friends, you have a much better chance of survival with a group around you.

Zebras mingle together for safety.

Not all grassland animals eat plants. There are many predators, such as the **cheetah**.

Giraffe

Fire!

Lightning strikes can cause the dry grasslands to catch fire. Wildfires are actually a good thing: They burn away old grass, making room for new grass to grow. In a fire, most animals and birds run, burrow, or fly away, but many insects die.

The **fork-tailed drongo** loves a good wildfire. This blue-black bird waits around to eat the barbecued insects.

Browsing animals, such as the **impala** and the **elephant**, reach up to eat the leaves and bark from plants.

Grazing animals, such as the **gazelle**, bend down to eat grass. There are more than 10,000 species of grass!

Savanna Spotlight: Giraffe

The **giraffe** is the tallest animal on Earth. Some male giraffes can be 20 feet (6 m) tall. That's as tall as a two-story house! A giraffe's neck is as tall as a man (6 feet/1.8 m), yet both giraffes and humans have just seven bones in their necks. Giraffes eat the leaves off tall, umbrella-shaped acacia trees. A giraffe's tongue and mouth are coated with mucus to protect them from the acacia thorns. Giraffes can go for days without water. Drinking water is tricky. A giraffe must spread its long legs and bend awkwardly to reach its head to the ground. Drinking is also dangerous, since the giraffe can't see its enemies in this position. When standing up, giraffes can see lions, spotted hyenas, and wild dogs from far away and get a running start. A giraffe can run faster than a horse. Giraffes also tend to sleep standing up. A giraffe only sleeps about five minutes at a time and only thirty minutes total a day. Why do they sleep so little? Giraffes are prey, and if they laid down on the open savanna for some serious shut-eye, they would instantly be another animal's midnight snack.

Wow!
FACT

The **springbok** likes to jump straight up into the air over and over again. When a group of springbok jump together, it's called pronking.

Mind the Mound

Termites build enormous mounds out of soil, saliva, and their own poop. They dig tunnels under the mounds, and up to two million termites live inside.

Termite

Different Name, Same Thing

Grasslands go by different names around the world. Some have different climates and grasses of different heights. The height of the grass depends on the amount of rain. The more rain, the higher the grass.

North America = prairie

South America = pampas

Africa = savanna or veld

Europe and Asia = steppe

Australia = bush or downs

73

Tundra

Brrrrrr . . . the tundra is the coldest biome. How cold? The lowest recorded temperature was -126.4° F (-88°C)! There are two types of tundra: **arctic** and **alpine**. Arctic tundra is found in the Arctic Circle by the North Pole and in the Antarctic Circle by the South Pole. Alpine tundra is found on high mountaintops throughout the world.

The tundra has only two seasons—winter and summer. Winters are long and cold, and summers are short and cool. It never gets hot on the tundra.

The tundra is often called a "cold desert." The temperature doesn't warm up much, so water stays frozen and plants and animals can't drink it. Frozen water doesn't evaporate, so little snow or rain falls.

Snowy owl

Caribou

Arctic ground squirrel

Moss campion

Short and Scrubby

"Tundra" is a Russian word that means "treeless area." Tree roots can't grow down into the permafrost, so only small, scrubby plants and shrubs with short roots can survive.

Many flowering plants, such as the **arctic poppy**, have a hairy stem to keep warm.

Oily Fur

The **polar bear** is the largest animal on the tundra. The polar bear's coat is very oily. The oil makes its fur waterproof, keeping the polar bear warm as it hunts for seals.

Hoofing It

Caribou are members of the deer family. The name "caribou" comes from a Native American word that means "snow shoveler." The caribou scrapes away snow with its hooves to find lichen to eat. In parts of Asia and Europe, caribou are called **reindeer**.

Permafrost

In the winter, all the soil in the tundra is frozen. In the summer, the top layer thaws, or warms up, but the bottom layer stays frozen. This frozen soil is called **permafrost**. Water cannot drain through permafrost, so bogs and marshes form on top of the soil in the warmer months.

Hollow Hair

A **musk ox** has a lot of long, thick hair that hangs almost to the ground. Each strand of its hair is hollow. Warm air gets trapped inside the hollow hair, keeping the musk ox toasty warm.

Smaller Is Better

The **arctic fox** has short ears, short legs, and a compact body, so less skin is exposed to the frigid air.

The Fat Layer

Many animals, such as walruses, whales, and narwhals, have a layer of fat called **blubber**. Blubber acts as an insulator. This means it holds in body heat, even when the animal is swimming in bone-chilling water.

Try It: The Blubber Glove

Find out if blubber really does keep you warmer.

You need:

Large bowl of cold water	Spatula
About 12 ice cubes	Gallon-sized plastic bag
Stopwatch	1–2 cups of shortening (such as Crisco)
Paper and pencil	Quart-sized plastic bag
Old sock, mitten, or glove	Duct tape

To do:

1. Place the ice cubes in the bowl of cold water.
2. As you submerge one of your bare hands into the bowl of ice water, have your partner start the stopwatch. When your hand starts to hurt or sting, pull it out. Write down how long your hand was in the water.
3. Now put a glove, mitten, or sock over your hand. Submerge your hand into the water again, and time how long before it starts to hurt. Write down this time.
4. Using the spatula, fill the gallon-sized plastic bag with the shortening. Shortening is a kind of fat that will act like blubber.
5. Place your hand inside the quart-sized plastic bag to keep it from getting greasy. Then place it inside the bigger bag of shortening. Mold the shortening around your bagged hand, adding more if necessary. Have your partner duct tape the bag around your wrist, so water will not get inside.
7. Submerge your "blubber glove" into the ice water and time how long before it starts to hurt. Write down this time.
8. Which way were you able to stay in the ice water the longest?

Spotlight: Penguins

Penguins are birds, but they can't fly. They use their wings as flippers to swim and dive in the water. They live only in the southern half of the world. Penguins eat krill, fish, or squid. Their enemy is the leopard seal. Once a year, penguins gather in an enormous group called a rookery to find a mate, build a nest, and lay eggs.

There are seventeen different kinds of penguins. Here are nine.

45 in. 114 cm									
36 in. 91 cm									
27 in. 69 cm									
18 in. 46 cm									
9 in. 23 cm									

The **Emperor penguin** is the biggest.

The **fairy penguin** is the smallest.

The **Adélie penguin** has a red bill.

The **gentoo penguin** is the fastest penguin swimmer.

The **rockhopper penguin** hops along the rocky shores.

The **macaroni penguin** has yellow feathers on its head.

The **chinstrap penguin** has a strip of black feathers from ear to ear.

The **Galapagos penguin** is the most northern-living.

The **king penguin** can live for 20 years.

What's on the Menu?

When's the last time you were hungry?

Every animal and plant gets hungry. We all need food for energy and to grow. Without food, all living things die.

Producers can make their own food. Plants are producers. They change the Sun's energy into food in a process called **photosynthesis**.

Consumers cannot make their own food. Animals are consumers. They must eat plants or other animals.

Scavengers don't hunt. They eat animals that are already dead, called **carrion**. They have sharp beaks or teeth to rip apart the dead animals. Vultures, hyenas, and gulls are all scavengers. Termites and earthworms are scavengers that feed on dead plants.

Decomposers are consumers that eat dead animals and waste. Without them, dead animals and plants would pile up in huge heaps. Earthworms, fungi, and beetles are decomposers.

Griffon vulture

So what did you eat when you were hungry?
A salad? A burger? Both? Plants and animals are placed into group by the types of food they eat.

Herbivores
Animals that eat plants are **herbivores**. Herbivores have back teeth called **molars** that grind up plants and grasses.

Carnivores
Animals that eat other animals are **carnivores**. Carnivores are hunters. They have four sharp, pointed teeth called **canines**. Canines help tear the skin and flesh of other animals.

The **grey wolf** hunts deer, elk, caribou, and rabbits. They live and travel in packs of seven or eight wolves. The leader of the pack is called the alpha wolf.

Omnivores
Animals that eat both plants and animals are **omnivores**. Omnivores have canines and molars.

A **raccoon** eats nuts and berries. It also eats frogs, mice, eggs, and earthworms. A raccoon is very clever. It can open trash cans and doors when it is looking for food.

Moose comes from a Native American word that means "twig eater." A full-grown moose can eat 50 pounds (23 kg) of twigs and shrubs a day. If a moose wants to eat a plant that is low to the ground, it has to kneel down.

Food Chains and Webs

All living things are connected. A **food chain** shows how energy is passed from one living thing to another.

HOW A FOOD CHAIN WORKS:

1 Plants trap the energy from the Sun for food.
2 Herbivores eat these plants. Herbivores get their energy from the plants.
3 Carnivores eat the herbivores. Carnivores get their energy from eating other animals.
4 When a plant or animal dies, the decomposers come in. The decomposer breaks down the dead plants and animals. It releases the nutrients back into the soil. Plants use these nutrients plus energy from the Sun to grow.

Sun

Plant

1

Herbivore

2

Carnivore

3

Carnivore

Decomposer

4

When an animal in one food chain eats an animal in another food chain, a **food web** is created. A food web shows what eats what in a community or ecosystem. If one link is taken away, the web is broken. If there were no carnivores, there would be too many herbivores and not enough producers to feed them. Predators and prey are both very important to keep nature in balance.

Every ecosystem has its own food web. Here is an ocean food web to give you an idea how it works.

Ocean Food Web

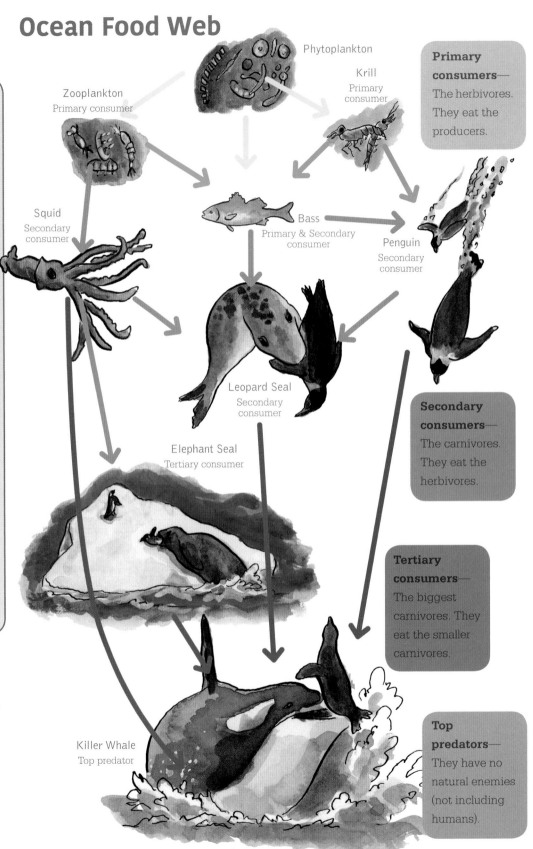

Phytoplankton

Krill
Primary consumer

Zooplankton
Primary consumer

Squid
Secondary consumer

Bass
Primary & Secondary consumer

Penguin
Secondary consumer

Leopard Seal
Secondary consumer

Elephant Seal
Tertiary consumer

Killer Whale
Top predator

Primary consumers— The herbivores. They eat the producers.

Secondary consumers— The carnivores. They eat the herbivores.

Tertiary consumers— The biggest carnivores. They eat the smaller carnivores.

Top predators— They have no natural enemies (not including humans).

Run and Hide: Camouflage and Other Defenses

Surviving in the natural world isn't easy. It's basically: Eat or be eaten. What do you do if a predator is coming after you? Some animals run fast, some hide, some make loud noise, and some fight back.

In Disguise

Many animals have developed a special color or shape that lets them blend in with their habitat. This is called **camouflage**. Camouflage works like a disguise. The idea is if you can't see me, you can't eat me—or if you can't see me, then I have a better chance of eating you.

The **leafy seadragon** looks like seaweed.

All for One

Staying in a group is sometimes the best way to hide. A group of **zebras** looks like a blur of black-and-white stripes to a lion. It's hard to pick out just one zebra. When not in a group, zebras like to stand in pairs. They stand facing in different directions to look out for predators.

The **arctic hare's** thick white fur keeps it hidden in the snow.

The **giant walking stick** looks like a tall blade of grass.

Tricked You!

If grabbed by a predator, the **opossum** goes completely limp, as if it were dead. An opossum can play dead for hours. Many predators only like living prey, so they drop the opossum and walk away.

If All Else Fails, Hide!

An **armadillo** curls up into a ball inside its hard armor. The predator can't get in and eventually gives up. Its name means "little armored one" in Spanish.

A Little Liquid Goes a Long Way!

A **skunk** squirts a horrible-smelling liquid from under its tail.

A **llama** spits saliva from its mouth or a bad-smelling liquid from its stomach. Llamas have great aim!

A **horned lizard** squirts blood from the corners of its eyes.

Eye See You

Many **butterflies** have false eyes. They have large spots on their wings that look like eyes. These fake eyes frighten off birds.

Hide-and-Seek

The **chuckwalla** is an iguana that can puff up like a balloon. When it is chased by a predator, it will squeeze into a crack in a desert rock and fill its lungs with air to inflate its body. Now the predator can't pull the chuckwalla out of its hiding spot.

Get the Point

When threatened, the **porcupine** turns its back to its enemy. It raises its sharp quills, making it look twice as big. Then it rattles them, making a loud noise. Finally it releases some quills—right into its enemy. The quills have a barbed tip, so they are painful to pull out. An adult porcupine can have 30,000 quills!

Run for It!

One of the best defenses is speed. A **squirrel** will run in a zigzag pattern to confuse the predator chasing it.

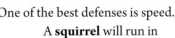

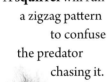

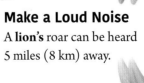

I'm Bigger Than You (Sort Of)

The **frilled lizard** has a flap of skin around its head that expands to make it look bigger and scarier to other animals.

Make a Loud Noise

A **lion's** roar can be heard 5 miles (8 km) away.

I've Got Your Back: Partners and Parasites

Does your friend protect you from bullies? Or does he or she steal your lunch and eat it, leaving you hungry? Just like with humans, animals and plants pair up with other animals and plants—some in helpful ways and some in not-so-helpful ways. When an animal pairs up with an animal of a different species or plant it's called a **symbiotic relationship**. "Symbiosis" comes from a Greek word that means "living together."

There are different kinds of symbiotic relationships:

Mutualism

Each species gains something from the relationship.
Oxpeckers are small birds that eat blood-sucking parasites off oxen (also off hippopotamuses, rhinoceroses, and zebras). The oxpeckers enjoy their meal, and the oxen end up with fewer bug bites.

The **spiny lobster** has figured out a great roommate system. **Octopuses** like to eat spiny lobsters. But **moray eels** eat octopuses. Spiny lobsters try to live in a den with moray eels. When an octopus goes to eat the lobster, the moray eel acts as its bodyguard and eats the octopus first.

Clownfish are coated in mucus that protects them from the painful stings of the **sea anemone**. The clownfish swims among the anemone's tentacles and lures bigger fish over to it. But the big fish are in for a huge surprise. The sea anemone stings and kills the big fish—and shares the leftovers of its meal with the clownfish.

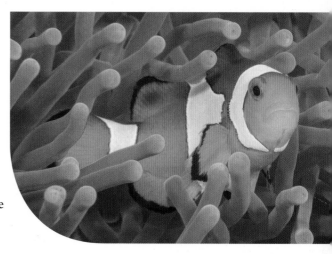

Parasitism

*One species gains and one is harmed. The one that gains is called the **parasite**. The one that is harmed is called the **host**. The host is usually larger than the parasite.*

A **tick** is an arachnid, and it feeds on the blood of other animals, such as **deer.** When it sucks on a deer's blood, the tick's own body inflates like a balloon. Then it falls off and molts and waits for a new host. Some ticks carry diseases, such as Lyme disease, which harm the host.

Mistletoe is a parasite. It digs its roots in **oak trees** or **apple trees** and sucks up all their nutrients. The tree doesn't die, but its branches don't grow correctly.

Commensalism

One species benefits while the other is neither helped nor harmed.

The **remora** attaches itself to the underside of a **shark** with suckers it has on its head. Since no one wants to mess with the shark, the remora is protected. The shark doesn't mind having a ride-along guest.

The long-legged **egret** lives in dry grasslands and eats insects. It likes to live with **buffalo**, because the buffalo's heavy hooves stir up the insects from the ground, making catching them a lot easier.

The **epiphytic orchid** lives in a fork or branch of a tall tree in a tropical forest. Way up high, the orchids are less likely to be eaten by animals on the ground and can find more sunlight. The orchid does not get any nutrition from the tree. All its nutrients come from sun and rain.

Fleas jump onto **dogs** and **cats.** They burrow into their fur to bite their skin and suck their flood. The fleas get food and a warm home. The dogs and cats get a horrible itch.

81

Group Living

The bell rings for recess. Some kids play kickball. Some kids climb the jungle gym. And some talk together. Whether 20 kids or only two kids, each group forms because the people in it have something in common. For animals, group living isn't all about fun—although puppies do love to play. Living in a group protects animals from predators and makes finding food easier.

Husky puppies

Protect and Defend

A group of African **lions** is called a pride. A pride usually has at least a one male lion, two or three lionesses, and their cubs. Each pride has its own territory, or area, that it defends. The lion's job is to protect the pride from predators. The lionesses hunt for food together. They use teamwork to trap and kill other animals. Even though the lioness catches the food, the lion always eats first, then the lionesses, and finally the cubs.

Bossy

A group of **chickens** is called a flock. Each flock has a "pecking order." This means that certain chickens are the "bosses." All the other chickens must do what the boss says. The chickens at the top of the pecking order can peck chickens that are lower than them, and the lower chickens can't peck back. The top chickens eat first and sleep in the better spots. Every chicken knows where he or she is ranked in the flock, and this helps keep peace. The chickens at the top of the pecking order are often the stronger or healthier chickens.

Clean-up Crew

A common group activity is grooming, or cleaning one another's bodies.

Japanese macaques, also known as snow monkeys, spend hours every day grooming each other.

Blow Your Horn

Bighorn sheep pick their group leader by horn size.

82

Animal Group Names

ANIMAL	GROUP NAME
Alligators	Congregation
Apes	Shrewdness
Baboons	Troop or congress
Badger	Cete
Bats	Colony or cauldron
Butterflies	Flutter or swarm
Cats	Clowder
Clams	Bed
Cobras	Quiver
Coral	Colony
Coyote	Pack
Crow	Murder
Donkeys	Drove
Eels	Swarm or bed
Emus	Mob
Falcons	Cast
Ferret	Business
Fish	School
Frogs	Army
Geese	Gaggle
Giraffes	Tower or herd
Gorillas	Band
Hamsters	Horde
Hyenas	Clan
Jaguars	Shadow
Lobsters	Risk
Mosquitoes	Scourge or swarm
Owls	Parliament
Porcupines	Prickle
Ravens	Unkindness
Salamanders	Maelstrom
Seahorses	Shoal
Sharks	Shiver
Storks	Mustering
Swans	Bevy
Toads	Knot
Turtles	Bale or dole
Water Buffalos	Pot
Whales	Pod or herd
Worms	Squirm
Zebras	Zeal or dazzle

The More, the Merrier

Some animal groups can be enormous!

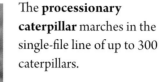

The **processionary caterpillar** marches in the single-file line of up to 300 caterpillars.

More than 50 **meerkats** can live together in a burrow.

Spiny lobsters walk very close together in a long row. The antennae of one touches the tail of another. Marching together makes it harder for predators to attack.

Listen Up

The **spotted hyena** is also called a laughing hyena, because it makes a loud, laugh-like sound to communicate with other members of its clan.

All Alone

Some animals would rather not share their food and living space with others. Animals that go it alone are called solitary animals.

A **leopard** lives alone, because it must eat 50 pounds (23 kg) of meat every few days, so it doesn't want to share what it hunts.

Stay, Go, or Sleep?

In places where winter grows cold and it is hard to find food, animals have three choices: stay active, hibernate, or migrate.

The Big Sleep

Hibernate means to sleep through the winter. While many animals sleep, their body temperature and heart rate slow, so they do not use much stored energy.

There are two kinds of hibernators: true hibernators and deep sleepers.

True Hibernators

True hibernators shut down their body's functions. They are barely alive. Their bodies become very cold. Their heartbeat and breathing slow down. They are very hard to wake up once asleep.

Wood frogs go to sleep in the winter and their bodies freeze like ice cubes. When spring comes, they thaw out and wake up.

A **woodchuck** loses half its body weight while it snoozes through the snow.

Deep Sleepers

Deep sleepers wake up and then go back to sleep throughout the winter. Their body temperature only lowers a little.

Throughout summer, a **black bear** eats as much as possible (that is where we get the phrase "hungry as a bear") to store fat. As the bear sleeps through the winter, all this fat will keep it alive. If a sleeping bear becomes aware of a predator, it can wake up to defend itself.

A **skunk** will wander around on warmer days.

Hideout

If you are asleep, you are easy pickings for predators. The key is to sleep where no one will find you. **Brown bats** hibernate in caves, the **queen bumblebee** hibernates in a hole in the soil, a **snail** snuggles into its shell, and **garter snakes** take a group nap in hollow logs.

See You Later

Migration is when animals move from one area or climate to another. They travel through water, in the air, or over land. Animals migrate to search for food, find warmer weather, or to have their babies. Migrating is often dangerous.

Birds are lucky. They can flap their wings and fly off to someplace better. Most other animals can't move from place to place so easily.

The **Canada goose** migrates in a group. They fly in a V-shape. In this shape, they can cut through the air more easily.

Born at Home

Salmon live in the ocean, but they return to the same freshwater river where they were born to have their babies. Salmon must swim upstream to get back home. They use their sense of smell and taste to guide their way there and back.

Danger Ahead

Many animals migrate in groups to stay safe from predators.

On Christmas Island, off the coast of Australia, millions of **red crabs** migrate every year. They walk 5 miles (8 km) from the forest to the ocean. The one-week trek is filled with danger. They must cross roads filled with cars. They climb down 40-foot (64-m) cliffs in the hot sun. They travel to the ocean to lay their eggs in the water. After they lay their eggs, they return to the forest. They always travel together and always take the same route there and back.

Gray whales swim every winter from the ocean around Alaska to the warm waters off Mexico—up to 13,700 miles (22,000 km) round trip. Along the way, they must dodge oil tankers and pollution and stay away from predators, such as orcas. Monterey Bay, California, is sometimes called "Ambush Alley." Orcas wait here for the migrating gray whales and attack them in the open waters of the bay. Every spring, the gray whales return north to look for food. On the long journey, they don't eat and live off their body fat.

Rest Stop

Many migrating animals make stops along the way.

Sandhill cranes always stop to eat at the Platte River in Nebraska on their round trip from Canada to the southern United States.

Asking for Directions

If someone dropped you thousands of miles from your house and told you to find you way home, could you? Migrating animals don't have maps. They don't have GPS or cell phones. They use their senses to remember where to go. Some animals use magnetic fields. Some use echoes. Some use the sun, moon, and stars.

Elephants migrate largely by sight. The oldest female elephant, or matriarch, remembers landmarks such as rivers and mountain ranges, and she leads the other elephants in the herd.

Animal Babies!

Starfish

In order for a species to survive, it must reproduce, or have babies. Some simple animals, such as the **starfish** and **sea anemones**, can reproduce by themselves. How do they do that? They split off part of their body and make a copy of themselves.

But most animals must find a partner of the opposite sex in order to have babies. Some animals give birth to live babies. Others lay eggs.

Live Birth

Most mammal babies grow inside their mother until they are born. This makes mammals different from other animals. Humans are inside their mother's wombs for nine months. **Alpacas** are inside for 12 months. And **elephants** stay in for 22 months—that's almost two years! Mammal babies drink milk from their mothers until they are old enough to eat other food.

Laying Eggs

Birds, reptiles, fish, and amphibians lay eggs. Bird eggs have a hard outer shell. Amphibian and fish eggs are soft and jelly-like. Some reptiles, such as snakes and turtles, have eggs with soft, leathery shells. Other reptiles, such as crocodiles and geckos, have eggs with a hard shell.

Chicken eggs and chicks

The number of eggs a bird lays at one time is called the **clutch size**. Some birds, such as the **albatross**, lay only one egg. Some birds, such as the **bobwhite**, lay 20 eggs. Some fish, such as the **Atlantic cod**, lay millions of eggs each season.

Many birds feed their young by chewing and then regurgitating (spitting up) their food and placing it in their chicks' mouths.

Some Dads Help

A **seahorse** father carries the eggs in a pouch on its stomach.

After the mother **emperor penguin** lays an egg, the father holds it on his feet. He covers it with a flap of skin to keep it warm. Mom and dad stand side by side for two months without eating or drinking until their chick hatches.

It's a Twin Thing

A lot of animals give birth to twins. The twins are usually **fraternal**, not identical, meaning they don't look exactly alike. Some animals that commonly have twins are: dogs, bears, polar bears, New World monkeys, deer, hyenas, and giant pandas. **Nine-banded armadillos** are usually born as identical quadruplets.

Polar bears and her cubs

A baby deer is a **fawn**.

Puppies can't see or walk when they are first born.

Do You Know Your Baby Animal Names?

ANIMAL	YOUNG		ANIMAL	YOUNG
Antelope	Calf		Horse	Foal / colt / filly
Bear	Cub		Kangaroo	Joey
Beaver	Kit		Owl	Owlet
Bobcat	Kitten		Peafowl	Peachick
Camel	Calf		Pig	Piglet
Caribou	Fawn		Pigeon	Squab / squeaker
Cow	Calf		Platypus	Puggle
Coyote	Pup		Rabbit	Kitten
Deer	Fawn		Rat	Pup
Eagle	Eaglet		Rhinoceros	Calf
Eel	Elver		Salmon	Parr / smolt
Elephant	Calf		Seal	Calf / pup
Ferret	Kit		Sheep	Lamb
Fish	Fry		Spider	Spiderling
Frog	Tadpole		Swan	Cygnet
Goat	Kid		Whale	Calf
Goose	Gosling		Zebra	Foal

A baby goat is a **kid**.

A baby owl is an **owlet**.

A baby rabbit is a **kitten**.

Protecting the Young

Babies are easy prey for predators, since they can't move as fast and haven't mastered the art of hiding. Mom or dad must often act as bodyguard.

To protect her babies from lions and hyenas, a **cheetah** mom moves her cubs to new hideouts every few days.

Anteater babies ride piggyback on mom to get safely from place to place.

All Gone . . . and Going

Extinct = gone forever

Endangered = in danger of becoming extinct

Today, a plant or animal species disappeared from Earth.

One disappeared yesterday, too.

And the day before that.

And the day before that.

At least 200 disappeared this year.

Where did they go? "Disappeared" makes it sound as if they were transported to another planet. The truth isn't so magical: Every single plant or animal in a species died. They will never be seen again. The scientific word for this is **extinction**. For every species that is alive today, about 1,000 have gone extinct. And many others are **endangered**, or on their way to disappearing forever.

Who's to Blame?

Mostly humans. According to *Science* magazine, extinction rates are 1,000 times higher than they would be if people weren't in the picture.

Scientists use the word "HIPPO" to help remember the causes of extinction.

= Habitat Loss

A habitat is an animal's home. If an animal loses its home and can't find a safe place to live, it will often die. Most habitat loss is caused by humans. We build roads, cities, and dams. In China, bamboo forests were cut down to make room for buildings and farms. The **giant panda** eats a lot of bamboo—28 pounds (8 kg) a day—and the pandas couldn't find enough food. They began to die. Giant pandas only have one or two babies during their lifetime, so not enough babies were born to replace the dying pandas. Today, only about 1,600 pandas are left in China's forests.

=Invasive Species

An invasive species is a plant or animal from another part of the world that doesn't belong where it now is. **Purple loosestrife** is a pretty flowering plant. It was brought to the United States from Europe in the early 1800s to decorate a garden. But one purple loosestrife plant can make up to 300,000 seeds in one year, so the plant grew and spread. In Europe, 100 different insects eat it and stop it from growing. But in the United States, there are no insects that eat it, so it grows out of control. This invasive plant crowds out and kills other wetland plants.

= Population Growth

There are more than 7 billion people living on Earth today—and the United Nations predicts our population may increase to 16 billion by the end of the 21st century! The Earth has a limited amount of space, food, and water. The more that people use, the less is left for animals and plants. Scientists have shown that population growth also worsens global climate change.

= Pollution

Pollution is caused by humans. Garbage, human waste, runoff from factories, and weed killers poison lake and river water and kill fish. Predators that eat the poisoned fish also get sick and die. Oil spills from large tankers in the oceans make it difficult for birds to fly and fish to swim. **Leatherback turtles**, **seals**, and **whales** often mistake plastic bags thrown into the water for jellyfish. The bags block their digestive systems and kill them.

= Over-hunting

People kill many animals to eat them. If these animals do not have a large population to begin with, they can be quickly hunted to extinction. Animals are also hunted for body parts, such as fur, feathers, or horns. In Africa, the **elephant** was hunted for its ivory tusks. The elephant population dropped from millions to a few hundred thousand. In China, the **tiger** was hunted for its valuable fur and for its bones, which were ground up and used for medicine. The tiger is now an endangered species. Some people shoot animals as a sport. Last year, an American man paid money to go on a "big game hunt" in Africa. He killed a **lion** named Cecil that had been tricked to come out of a protected area. Cecil isn't alone. Hundreds of African lions are killed every year "for fun."

What's Global Climate Change?

The Earth is getting warmer. Scientists have the research to prove this. They've also shown that the climate is changing much faster than ever before. Why? It has to do with a gas, carbon dioxide. Carbon dioxide traps the Earth's heat, and that's normally a good thing. It keeps our oceans from freezing and allows us to have life on Earth. But now there is too much carbon dioxide in our air from humans burning coal and using a lot of gasoline. As the levels of carbon dioxide go up, the climate grows warmer. The ice in the Arctic melts and ocean levels rise. Land starts to flood and weather patterns change. All this messes with food webs—and when one link goes extinct, the whole web is in jeopardy.

Endangered

The **addax** is a type of antelope. It used to be found all over the Sahara Desert in Africa, but habitat loss and overhunting has made it endangered.

The **aye-aye** is the world's largest nocturnal primate. Destruction of its forest home in Madagascar has made it endangered. Also, because of the local superstition that the gremlin-looking aye-aye brings death, people have been killing them for years.

Addax

The **black rhinoceros** has been killed for years for its horn. There are only a few left in Western Africa. The name "**rhinoceros**" comes from two Greek words meaning "nose" and "horn." Some people believe a rhino's horn has magical powers.

The **green sea turtle** lays its eggs on sandy beaches, but when the beaches were paved over to build hotels for tourists the green turtles were left with nowhere to go. The turtles have also been overhunted for their meat and eggs.

Green sea turtle

The **Lear's macaw** is a large blue bird from Brazil. One macaw can eat up to 350 nuts a day from a certain type of palm tree. When these trees were cut down for farming and to build roads, the Lear's macaw died in huge numbers.

Lear's macaw

Aye-aye

How Does an Animal Species Get Called "Endangered"?

In the United States, if someone—anyone—is worried that a species is close to extinction, he or she can send a letter to the U.S. Fish and Wildlife Service. The Service investigates. If they find enough scientific evidence, that species will go on the official endangered list. Worldwide, there are about 16,000 animals listed as endangered.

Go to http://www.fws.gov/endangered/ if you want to read the list.

In many countries, it is a crime to kill or hurt an endangered animal. Endangered animals are often placed in **wildlife** or **nature preserves**, large areas of land where animals and their habitat are protected. Building of houses and roads is limited or not allowed here. Hunting is also limited or illegal. This helps scientists keep the species alive.

Success Story!

In 1941, there were just **41 whooping cranes** left in the world. They'd come close to extinction from overhunting for their feathers and habitat loss. But scientists and naturalists got together to help save the tall birds. Today, there are over 600 whooping cranes!

Extinct

The **dodo**, which went extinct in 1681, was a flightless bird that lived on the island of Mauritius near India. It had no natural predators there. Humans arrived on the island in the early 1500's. They brought dogs with them. They also, without meaning to, brought rats in their boats. The dogs and rats ate the dodo's eggs, and the people hunted the dodos for food.

The **Tasmanian tiger**, which went extinct in 1936, looked like a tiger-wolf mash-up. It had a marsupial pouch, like a kangaroo. When people settled on the island of Tasmania in the 1800s, they over-hunted the Tasmanian tiger.

The **quagga**, which went extinct in 1883, was a type of zebra that lived on the grasslands of South Africa. It was hunted for meat and for its hide to make leather. Farmers killed it, so it wouldn't compete with their sheep and goats to eat the grass. Its name came from the sound it made.

The **Eskimo curlew**, which went extinct in 1981, was a shorebird with a long, thin beak. Two hundred years ago, there were millions of them migrating between Canada and South America. Along the way, the Eskimo curlew used to eat the Rocky Mountain grasshopper. After that grasshopper went extinct and hunting of the bird rose, the Eskimo curlew also became extinct.

Get to Know: Rachel Carson
(1907–1964)

Rachel Carson was an environmentalist and writer who warned the world of the dangers of chemicals to life on Earth. Rachel grew up on a farm in Pennsylvania. She got a degree in zoology from Johns Hopkins University. Then she took a job with the U.S. Bureau of Fisheries and then became a marine biologist. She wrote articles and books about the ocean and nature. In 1962, she published her fourth book about nature, called *Silent Spring*. The book begins in a regular town in America, but sickness and death overtake the town. The birds, the cows, the trees and flowers, and even the humans die. Carson explained that the cause was pesticides, chemicals used by farmers to kill insects. These chemicals had seeped into the soil and the water and were killing every part of the worldwide food web. People were outraged, and they protested in Washington, D.C. Carson had opened their eyes to the dangers of environmental pollution. Carson died in 1964 of cancer, but in 1970, the Environmental Protection Agency (EPA) was formed to stop pollution of water, air, and soil.

All About Natural History Museums

How the Museums Came to Be

Back in the 1500s, European explorers and scientists began voyaging to faraway continents. They brought back strange plants and animals that didn't grow or live in Europe. People were curious. They wanted to see these amazing oddities for themselves. The scientists displayed their discoveries in "cabinets of curiosities" and showed them to very wealthy people and royalty.

They weren't actual cabinets, but rooms filled with wondrous items. Preserved animals. Huge tusks and horns. Exotic shells. The idea was to amaze the viewer, showing him something he'd never seen before—the stranger, the better. The real and the fantastic shared the same space. Collectors liked to show off fake unicorn horns.

By the 1700s, people realized that nature itself was strange and beautiful and wondrous. More importance was placed on collecting, describing, organizing, and displaying. Wealthy people liked to have their own large collections of plants, rocks, shells, and animals to show off. Eventually these collectors died from old age. What should be done with the hundreds of thousands of natural history specimens they left behind? And that's how the first natural history museum was started in London, England. Soon others followed around the world. Museums weren't just for the wealthy anymore. Everyone could visit!

What's Inside a Natural History Museum?

Everything that's in this book—and much, much more! And it's all life-sized and amazing. There are hundreds of exhibits for you to explore and lots of hands-on learning. Stand next to a real dinosaur skeleton. Stand under a gigantic blue whale. Stand inside a swarm of butterflies. Stand beside an enormous rare diamond. All the treasures of our natural world are waiting for you at a museum.

Why Are These Museums Important?

Museums tell us about every organism that ever lived or is living on Earth. They tell us about Earth itself—and how it is changing. Museums keep records of every type of plant, animal, gem, fungus, and fossil that has been discovered—and they show us up-close exactly how they looked and lived. These scientific records and actual specimens help scientists better understand biodiversity and environmental changes and help them predict the future of our natural world.

There are about 200 natural history museums all over the United States. Some are so enormous that they take up full city blocks, while others are just one room. No matter the size, natural history museums let us see and touch parts of the same big story—the story of life on Earth.

Fiberglass model of a blue whale at the American Museum of Natural History in New York.

Natural History Museums in the United States and Canada

California Academy of Sciences
San Francisco, California

Natural History Museum of Los Angeles County
Los Angeles, California

University of Colorado Museum of Natural History
Boulder, Colorado

Peabody Museum of Natural History at Yale University New Haven, Connecticut

Florida Museum of Natural History
Gainesville, Florida

Fernbank Museum of Natural History
Atlanta, Georgia

Bishop Museum
Honolulu, Hawaii

Field Museum
Chicago, Illinois

Burpee Museum of Natural History
Rockford, Illinois

Cape Cod Museum of Natural History
Brewster, Massachusetts

University of Michigan Museum of Natural History
Ann Arbor, Michigan

Museum of the Rockies
Bozeman, Montana

American Museum of Natural History
New York, New York

Cleveland Museum of Natural History
Cleveland, Ohio

Sam Noble Oklahoma Museum of Natural History
Norman, Oklahoma

Academy of Natural Sciences
Philadelphia, Pennsylvania

Carnegie Museum of Natural History
Pittsburgh, Pennsylvania

Museum of Natural History and Planetarium
Providence, Rhode Island

Brazos Valley Museum of Natural History
Bryan, Texas

Perot Museum of Nature and Science
Dallas, Texas

Natural History Museum of Utah
Salt Lake City, Utah

Virginia Museum of Natural History
Martinsville, Virginia

Burke Museum of Natural History and Culture
Seattle, Washington

The Smithsonian Natural Museum of Natural History
Washington, D.C.

Wyoming Dinosaur Center
Thermopolis, Wyoming

Royal Tyrrell Museum of Paleontology
Alberta, Canada

Canadian Museum of Nature
Ottawa, Canada

Natural History Museums Around the World

The Western Australian Museum
Perth, Australia

Royal Belgian Institute of Natural Science
Brussels, Belgium

Beijing Museum of Natural History
Beijing, China

Zigong Dinosaur Museum
Zigong, China

The Natural History Museum of Denmark
Copenhagen, Denmark

The Natural History Museum
London, England

Museum für Naturkunde
Berlin, Germany

National Museum of Natural History
New Delhi, India

Museum of Natural History at the University of Florence
Florence, Italy

Natural History Museum of Zimbabwe
Bulawayo, Zimbabwe

The Natural History Museum in London, England

Index

adaptation, 25
Age of Dinosaurs, 19
algae, 53
alligators, 57
alpine tundra, 74
amphibians, 16, 17, 32, 50–51, 86
animals, 9, 11, 16–18, 22–26, 32–47, 50–57, 60–63, 66–79, 82–91
annelids, 32, 34
Anning, Mary, 21
ants, 39
apes, 68, 69
aphids, 38
apiology, 40
arachnids, 35, 37
arachnology, 35
Archaeopteryx, 17
arctic tundra, 74
arthropods, 32, 35–37
arthropology, 35
atmosphere, 10
Audubon, John James, 63

baby animals, 86–87
bacteria, 16, 30
bark (trees), 29
bats, 67
beavers, 53
bed bugs, 38
bees, 40
beetles, 39
biodiversity, 11
biology, 8
bioluminescence, 39
biomes, 33, 48–49, 52–53, 58–59, 64–65, 72–75
birds, 25, 32, 60–63, 86
bivalves, 43
blubber, 75
bonobos, 69
bony fish, 45
botany, 26
breathing, 44
bulbs, 26
butterflies, 41

caecilians, 50, 51
camels, 59
camouflage, 78–79
canines, 67, 76
carcinology, 35
carnivores, 76
carnivorous birds, 60, 61
carnivorous plants, 29
Carson, Rachel, 91
cartilaginous fish, 45
cells, 17
cephalopods, 43
chimpanzees, 69
chlorophyll, 27
cicadas, 38
Clark, Eugenie, 47
classes, 32
climate, 16, 33, 89
cnidarians, 43
cold-blooded animals, 32, 54
commensalism, 81
conchology, 42
cone shells, 42
coniferous forest, 33, 64
conifers, 64
consumers, 76
coral reefs, 48–49
crocodiles, 57
crustaceans, 35, 36
crystal, 13

Darwin, Charles, 24, 25
deciduous forest, 33, 64
decomposers, 76
deep sleepers, 84
defenses (animals), 78–79
defenses (plants), 29
descended, 24
deserts, 33, 58–59
dinosaurs, 16–20
dolphins, 71
dormant, 64
drupes, 28

Earth, 10–11, 16–18, 48
echinoderms, 43
ecosystem, 33
eggs, 20, 86
elephants, 67
endangered species, 88–90
entomology, 38
erosion, 15
evaporate, 58
evergreens, 64
evolution, 24–25
exoskeleton, 35
extinct, 18
extinction, 22, 88–91

family group names, 83
felines, 67
ferns, 27
fins (fish), 44
fish, 16, 17, 32, 44–47, 86
flies, 38
flowering plants, 17, 27, 29
food chains and webs, 76–77
forests, 33, 64–65
fossils, 18
fraternal twins, 87
freshwater biome, 33, 52–53
frogs, 50, 51
fungi, 31

gastropods, 42
gemology, 12
gems, 13
geodes, 13
geology, 12
germinate, 27
germs, 30
giraffes, 73
global climate change, 89
Goodall, Jane, 69
gorillas, 69
grasshoppers, 41
grasslands, 33, 72–73
group living, 82–83

habitat, 33
habitat loss, 88
herbivores, 76
herbs, 28
herpetology, 54
hibernation, 84
hominins, 18
Homo sapiens, 18
hoofed mammals, 70

Ice Age, 22–23
ichthyology, 44
igneous rocks, 15
iguanas, 56
insects, 35, 38–41
invasive species, 88
invertebrates, 32
Irwin, Steve, 57

jawless fish, 45
jointed legs, 35

lakes, 52
leaves, 28, 64
Leeuwenhoek, Antonie van, 31
life, 11
life timeline, 16–18
lift (birds), 62
liquid water, 11
living things, 9
lizards, 56

magma, 15
malacology, 42
mammals, 18, 32, 66–71, 86
manatees, 71
marine biology, 44
marine biome, 33, 48–49
marsupials, 66
meat-eating dinosaurs, 19, 20
meat-eating plants, 29
metamorphic rocks, 15
metamorphosis, 41, 50
meteorites, 13
microbes, 30
microbiology, 30

microlife, 30–31
microorganisms, 30
migration, 84, 85
mineralogy, 12
minerals, 12–15
mites, 37
Mohs, Friedrich, 14
Mohs' scale, 14–15
mold, 31
mollusks, 32, 42–43
molt, 35, 63
monkeys, 68
monotremes, 66
mosses, 27
mouth (rivers), 53
museums, 92–93
mushrooms, 31
mutualism, 80
mycology, 30
myriapods, 35, 36

natural history, 8
natural history museums, 92–93
natural selection, 25
naturalist, 9
nature preserves, 90
nest building, 61
newts, 51
non-flowering plants, 27

ocean biome, 33
ocean food web, 77
oceanographers, 48
oceans, 11, 48
omnivores, 76
orangutan, 69
ores, 13
organic matter, 28
organisms, 11
ornithology, 60
over-hunting, 89

paleontology, 19
parasites, 37
parasitism, 81
penguins, 75
permafrost, 74
photosynthesis, 27, 76
placental mammals, 66, 67
plankton, 48
plant-eating dinosaurs, 19
plants, 9, 11, 16–18, 26–29, 64–65,
   72, 74, 88, 89
Pliny the Elder, 9
pollen, 27, 40
pollination, 27
pollution, 89
polyps, 48
ponds, 52
population growth, 89
precious metals, 13
predators, 33
prehensile tail, 68
prehistoric animals, 22–23
prey, 33
primates, 68
primordial soup, 17
proboscis, 38
producers, 76
pterosaurs, 16, 21

rainforest, 65
raptors, 61
reproduction, 86
reptiles, 21, 32, 54–57, 86
rivers, 53
rocks, 12–15
rodents, 68

salamanders, 50, 51
salmonella, 30
sand dunes, 59
sauropods, 19
scales (fish), 45
scales (reptiles), 54
scavengers, 36, 76
seas, 48
sediment, 15

sedimentary rocks, 15
seed coat, 27
seeds, 26, 27
segmented body, 35
sharks, 47
sleeping, 46
slugs, 42
snails, 42
snakes, 55, 57
soil, 28
solar system, 10
species, 24, 25
spices, 28
spiders, 37
spinneret, 37
spiracles, 38
stink bugs, 38
stone fruits, 28
streams, 53
swim bladder, 44
symbiotic relationships, 80

talons, 60
temperate grasslands, 72
tentacles, 43
theropods, 20
toads, 50, 51
tortoises, 55
trees, 29, 64–65, 72
trenches (ocean), 48
tributary, 53
trilobites, 16
tropical grasslands, 72
tropical rainforest, 33, 65
true hibernators, 84
trunk (trees), 29
tubers, 26
tundra, 33, 74–75
Turner, Charles Henry, 39
turtles, 55
twin animals, 87

ungulates, 70

vertebrates, 32

warm-blooded animals, 32
water, 11, 59
water bugs, 38
water cycle, 11
water mammals, 71
waterfowl, 60
weather, 33
weathering, 15
weeds, 28
whales, 71
whelks, 42
wildlife preserves, 90
worms, 34

zoology, 66
zooxanthellae, 48

Size Comparisons

Sometimes it's tricky to create a picture in our mind of how tall or long something is if we are given only ruler measurements. This list of everyday objects and their sizes will help you better compare the animals and plants in this book.

A soda can
= 4.75 inches
(11.4 cm) tall

This book
= 10 inches
(26 cm) tall

Standard grocery cart
= 38 inches (96.5 cm) tall

Emperor penguin
= 4 feet (1.2 m) tall

Homo sapiens (people)
= 5–6$\frac{1}{2}$ feet (150–180 cm)

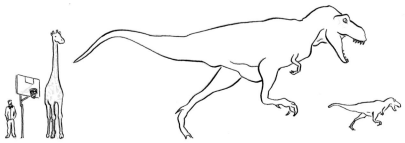

Regulation height
of a basketball hoop
= 10 feet (300 cm) tall

Giraffe
= 15–17 feet
(450–510 cm) tall

Tyrannosaurus Rex
= 20 feet (6 m) tall, 40 feet
(12 m) long

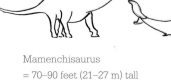

Yellow school bus
= 45 feet (13.7 m) long

Mamenchisaurus
= 70–90 feet (21–27 m) tall

Blue whale
= 80–100 feet (24–30 m) long